Reprints of Economic Classics

THE ECONOMIC HISTORY OF IRELAND
IN THE SEVENTEENTH CENTURY

Also by GEORGE O'BRIEN

In REPRINTS OF ECONOMIC CLASSICS

The Economic History of Ireland in the Eighteenth Century [1917]

The Economic History of Ireland from the Union to the Famine [1921]

An Essay on Medieval Economic Teaching [1920]

An Essay on the Economic Effects of the Reformation [1923]

£9.25

PRESSMARK
88198/330.9415/OBR

STRANMILLIS COLLEGE BELFAST
SN 0010406 X

UDS phone
17/4/80

THE
ECONOMIC HISTORY
OF IRELAND

IN THE SEVENTEENTH CENTURY

BY

GEORGE O'BRIEN

[1919]

AUGUSTUS M. KELLEY • PUBLISHERS
CLIFTON 1972

First Edition 1919
(Dublin: Maunsel & Company Ltd., 1919)

Reprinted 1972 by
Augustus M. Kelley Publishers
REPRINTS OF ECONOMIC CLASSICS
Clifton New Jersey 07012

By Arrangement With GEORGE O'BRIEN

. . . .

I S B N 0-678-00817-5
L C N 68-56555

. . . .

PRINTED IN THE UNITED STATES OF AMERICA
by SENTRY PRESS, NEW YORK, N. Y. 10013

TO
ALICE STOPFORD GREEN.

NOTE.

I WISH to express my gratitude to Mr. P. S. O'Hegarty, who was good enough to read and suggest some alterations in the manuscript of a section of the first chapter; to Father T. A. Finlay, S.J., for his great kindness in revising the manuscript, and in assisting me to correct the proofs; and to Mr. Arthur Cox, who also gave me much assistance with the proofs. The index was compiled by Mr. E. A. Phelps, to whom I am much obliged.

NOTE ON REFERENCES.

The following abbreviated references are employed throughout this book :—

Reference.	Title of Work referred to.
BAGWELL	Ireland under the Stuarts and during the Interregnum, by Richard Bagwell, London, 1909-16.
C.S.P. (IRE.)	Calendar of State Papers (Ireland).
C.S.P. (DOM.)	Calendar of State Papers (Domestic).
CARTE	Life of James, Duke of Ormond, by Carte, Oxford, 1851.
DUNLOP	Ireland under the Commonwealth, by Robert Dunlop, Manchester, 1913.
I.C.J.	Irish Commons Journal.
PRENDERGAST	The Cromwellian Settlement of Ireland, by John P. Prendergast, London, 1870.

References to Boate's *Ireland's Natural History* and to Davies' *Discoverie* are to the reprints of these works in *Tracts and Treatises relating to Ireland*; and references to Petty's *Political Anatomy* are to the first edition.

TABLE OF CONTENTS.

	Page
INTRODUCTION	1

Chapter

I. THE PERIOD OF CONSTRUCTION, 1603-1641 - 8

(1) Generally, p. 8. (2) The People and their Industrial Character, p. 12. (3) The Settlement of the Land System, p. 15. (4) The Cottiers, p. 33. (5) The State of Agriculture, p. 36. (6) Pasture and Tillage, p. 41. (7) The Woods, p. 43. (8) The Mines, p. 47. (9) Trade, p. 57. (10) The Woollen Industry, p. 68. (11) The Linen Industry, p. 75. (12) The Fishing Industry, p. 80. (13) Minor Industries, p. 83. (14) Public Finance, p. 85. (15) Coinage and Credit, p. 94.

II. THE PERIOD OF DESTRUCTION, 1641-1660 - 100

III. THE PERIOD OF RECONSTRUCTION, 1660-1689 - 116

(1) Generally, p. 116. (2) The People and their Industrial Character, p. 122. (3) The Land System, p. 127. (4) The Cottiers, p. 136. (5) The State of Agriculture, p. 142. (6) Pasture and Tillage, p. 144. (7) The Woods, p. 147. (8) The Mines, p. 148. (9) Trade, p. 150. (10) Industries in General, p. 173. (11) The Woollen Industry, p. 176. (12) The Linen Industry, p. 187. (13) The Fishing Industry, p. 193. (14) Minor Industries, p. 195. (15) Public Finance, p. 197. (16) Coinage and Credit, p. 203.

IV. THE PERIOD OF REDESTRUCTION, 1689-1700 - 211

APPENDICES.

Appendix		Page
I.	Tables of Prices in Dublin, 1599-1602	241
II.	Wages in Ireland in 1608 and 1640	242
III.	Extract from a Letter from Wentworth to the Lord Treasurer, dated 31st January, 1633, proposing a Monopoly of Salt	244
IV.	Ormond's Protest Against the Cattle Act	246
V.	Contemporary Accounts of Damage inflicted on Ireland by the Cattle Acts	248
VI.	Proposals made by Sir William Temple in his Essay on the Trade of Ireland for the Regulation of the Provision Trade	254
VII.	Sir William Temple's opinion of the Cattle Acts, from his Essay on the Trade of Ireland	256
VIII.	Memorandum urging the English Parliament to Repeal the Cattle Acts	257
IX.	An Act for the Advancement and Improvement of Trade, and for Encouragement and Increase of Shipping and Navigation	261
X.	Proposals of Thomas Sheridan on behalf of himself and others for Buying all the Wool of Ireland for 21 years, and for its transportation into England only, to commence from 1st May, 1674	265
XI.	Agreements in connection with the Woollen Manufacture at Callan	268
INDEX		271

THE ECONOMIC HISTORY OF IRELAND IN THE SEVENTEENTH CENTURY

INTRODUCTION.

WHILE studying the economic history of Ireland in the eighteenth century,[1] the writer of this book found his attention being perpetually directed to that of the seventeenth. The evils of the Irish land system, which assumed such terrible proportions in the later, flowed directly from the confiscation and degradation of the earlier century; the paralysis of industrial life under George II. and George III. was caused by English commercial legislation in the reigns of Charles II. and William III.; and the whole fabric of eighteenth century public finance, with its many merits and many abuses, was built on the settlement which followed the Restoration. Indeed, it became more and more apparent to the author that the eighteenth century but yielded the harvest of the crop which had been sown in the seventeenth, and that the phenomena witnessed in a study of the later period were but the effects of causes which began to operate in the earlier. The two centuries appeared to be so inextricably interwoven that a study of one by itself seemed doomed to be incomplete, just as an examination of the lower reaches of a river would be incomplete without some study of its source. It was this feeling of a work half done which induced the author to attempt this present volume.

[1] See *Economic History of Ireland in the Eighteenth Century*. Maunsel & Co., Dublin.

The economic history of Ireland in the seventeenth century may be divided into well-defined periods. The beginning of the century was marked by a complete breakdown of the whole economic life of the country, after which the industrial organization had to be built up anew from the foundations. The last quarter of the sixteenth century witnessed a devastating war in Ireland, in the course of which property of all kinds was deliberately and ruthlessly destroyed. It is not necessary for the purposes of this book to go into the vexed question of the type of social and economic life which English policy set itself to destroy; that is a matter on which competent historians have expressed strongly differing views; and it would be presumption for the present author to express any opinion on the subject without having made an independent investigation. It is sufficient to say that, whatever was the condition or quality of that earlier Irish agricultural and industrial organisation, whether it was, as some have suggested,[1] complex and highly developed, or, as others have asserted,[2] primitive and semi-barbarous, it was utterly and completely destroyed in the closing quarter of the sixteenth century.

The whole of Ireland was impoverished and laid waste. On the one hand, the Irish made it an essential part of their plan of defence to "keep their countries of purpose waste, uninhabited, as where nothing is, nothing can be got";[3] while, on the other hand, the English soldiers laid waste the districts through which they passed. "Along the track of Elizabeth's soldiers, houses, cornfields, orchards, fences, every token of a people's industry were laid 'handsmooth.'"[4] The Lord President of Munster having heard that there were fugitives lurking in certain parts of the province, "burnt all the houses and corn, taking great preys . . . . and

[1] *e.g.*, Mrs. Green's *Making of Ireland and its Undoing.*
[2] *e.g.*, Cunningham's *Growth of English Industry and Commerce.*
[3] Green, *Making of Ireland and its Undoing*, p. 113.
[4] Green, *op. cit.*, p. 90.

harassing the country, not leaving behind him man or beast, corn or cattle."¹ "The land itself," says Holinshead, "which before these wars was populous, well inhabited, and rich in all the good blessings of God—being plenteous of corn, full of cattle, well stored with fish and other good commodities—is now become so barren both of man and beast that, whoever did travel from the one end of all Munster, even from Waterford to the Head of Smerwick, would not meet any man, woman, or child, save in towns or cities; nor yet see any beasts but the very wolves, foxes, and the other like raving beasts."² "From Dingle to the Rock of Cashel," say the Four Masters, "not the lowing of a cow nor the voice of the ploughman was that year (1582) to be heard." "Your honour must conceive," writes Sir Francis Shane to Sir Robert Cecil in 1600, "that as long as the Irish shall be suffered never so little tillage with their cows, they will never quail."³ At the same time, "dispersed garrisons" were employed "keeping the women and children from ploughing, so that they must the next year starve."⁴ "Captain Blaney findeth the country all distracted and ready to starve, if they reap not the next crop, which I hope they shall not."⁵ The cutting down of green crops wherever they went was a deliberate part of the English campaign;⁶ a part of their operations which they conducted so successfully that the very cattle starved in the field.⁷ A whole generation afterwards, the destroyed buildings had not yet been replaced.⁸

If such a state of affairs had lasted for two or three years, it is possible that, when peace was restored, the damage done might have been repaired, and the land might have recuperated. In Ireland, however, this deliberate campaign of destruction lasted for upwards of twenty years, at the end of which the erstwhile prosperous countryside was a desert waste. In 1586 the grievous

---

¹ *Pacata Hibernia*, pp. 189-90. ² Holinshead, vol. vi., p. 459.
³ C.S.P., Ire., 1600-1, p. 196. ⁴ C.S.P., Ire., 1600-1, p. 206.
⁵ C.S.P., Ire., 1600-1, p. 247. ⁶ C.S.P., Ire., 1600-1, pp. 197, 355.
⁷ C.S.P., Ire., 1600-1, p. 86. ⁸ Gernon, *A Discourse of Ireland*, 1620.

decay of tillage and husbandry was deplored in the Irish Parliament;[1] and in 1600 the whole country was said to be a waste.[2] When Mountjoy made his entry into Cork, the citizens "entertained him with a show of plough-iron on both sides of the street, intimating thereby that the soldiers, by their exactions and rapine, had wasted the country, making all the ploughs to be idle which should have maintained it."[3] "Corn there is none," we read in 1600, "and but little grain to be cut in the country;"[4] and in that year the inhabitants of the former granary of Europe had to be supported on Dantzic rye and Newfoundland fish.[5] "The country is wasted and eaten up . . . spent to the bones."[6] "In the civil wars," wrote Fynes Moryson, "the great lords and inferior gentlemen laboured more to get new possessions for inheritance, than by husbandry and peopling of their own lands to increase their revenue."[7]

Of course, whatever industry and trade Ireland possessed suffered in this general devastation. To discuss how far the native Irish had succeeded in developing manufacturing industries and a foreign trade would be to embark on a highly controversial problem with which this book is not directly concerned; but it is perfectly clear that, as one of the aims of the English Government during the sixteenth century was to impoverish their enemies in Ireland, as elsewhere, an effort would be made to deprive them of whatever commerce they enjoyed. That such an attempt was, in fact, made, appears from the laws which forbade trading with a man of Irish name, or in an Irish fair, and also from the attacks made by English ships on trading vessels plying between Ireland and Spain.[8] It may be well to draw attention to the fact that these attacks on Irish commerce were of an utterly different nature from the attacks made in later centuries under the system which came to

---

[1] Green, *op. cit.*, p. 95.
[2] *A Treatise on Ireland*, by John Dymock, 1600; C. S. P. Carew, 1601-3, p. 423.
[3] C.S.P., Ire., 1606-8, p. xlix.   [4] C.S.P., Ire., 1600-1, p. 532.
[5] C.S.P., Ire., 1600-1, p. 414.   [6] C.S.P., Ire., 1600-1, pp. 66-83; 1603-6, p. 115.
[7] *Itinerary*, p. 221.   [8] Green, *op. cit.*, pp. 131-53.

be known as the Commercial Restraints. The aim of the English statesmen of the later seventeenth and eighteenth centuries was to damage the trade and prosperity of Ireland considered as a geographical entity; the statesmen of the sixteenth century, on the other hand, had no objection to the trade of Ireland flourishing, provided that the profits of that trade were reaped by the English interest in Ireland. Their aim was to divert the wealth of Ireland from one set of owners to another; from the Queen's enemies to her subjects; from those who would use it to assail her to those who would use it to defend her; just as they would have sought to paralyse the commerce of any county of England that fell into rebellion. However, the result of the trade policy of Elizabeth, together with the general destruction of wealth which took place in Ireland as incidents in the wars of her reign, was to cripple whatever Irish industry there had been, and to destroy Irish trade.[1]

The result of this destruction of industry, combined with the universal decay of agriculture, was disastrous in the extreme, and Ireland, during the first five years of the seventeenth century, was in a state of great poverty and want. The prices of provisions in Dublin were indicative of famine,[2] and the dispatches of the time are full of reference to the high prices of food.[3] "The estate of this poor country is grown most miserable, and there is a general dearth of all necessaries."[4] "Victuals are so scarce that it is thought most of the people will starve this year."[5] "There is an excessive dearth in the Pale; soldiers and handicraftsmen cannot live of their earnings."[6] In 1603, provisions of all kinds had to be imported from England, or the whole country would have starved.[7] During the early years of James's reign, Connacht and Ulster only produced enough cattle to keep the army for three months in the year.[8] In 1606 we read

[1] Green, *op. cit.*, pp. 165-6.
[2] Three lists of prices in the years 1599-1602 are printed in Appendix I.
[3] See *e.g.* C.S.P., Ire., 1601-3, pp. 279, 305. 562-3.   [4] *Cecil MSS.* (1602), vol. xii., p. 646.
[5] *Cecil MSS.* (1602), vol. xii., p. 650; see C.S.P , Ire., 1601-3, pp. 305, 560.
[6] C.S.P., Ire., 1601-3, p. 667.   [7] C.S.P., Ire., 1603-6., p. 118.
[8] C.S.P., Ire., 1611-14, p. 457.

of "the general waste and decay of that Kingdom which yields not that abundance that it did in former times;"[1] and in 1607 that "the whole country is waste, those who have escaped the sword having perished through famine," and that "the whole realm is more utterly depopulated and poor than ever before for many hundred years."[2]

Ireland, therefore, had to commence a new economic life at the beginning of the seventeenth century. In the following study of the economic history of Ireland during that century, the subject is divided into four chapters. The first chapter is concerned with the economic life of the country from 1603 to 1641, during which time efforts were being made, not without success, to settle the ownership and tenure of land on a new and stable plan, and to rebuild the shattered industries and foreign trade. These forty years constitute the *Period of Construction*. The second chapter deals with the twenty years following the outbreak of the Rebellion in 1641—a period characterized by the wholesale destruction of the population, the devastation and impoverishment of the land, the degradation of the old propertied class, and the ruin of all manufactures and commerce. These twenty years constitute the *Period of Destruction*. The third chapter is devoted to a consideration of the period between the Restoration and the Revolution, during which a new and more resolute effort was made to encourage industry, and the nature and direction of Irish foreign trade were changed by the operation of the first acts of the new English commercial policy. These thirty years constitute the *Period of Reconstruction*. Finally, the fourth chapter deals with the last ten years of the century, during which the foundations of the penal code were laid with the object of depriving the majority of the Irish people of all wealth and ambition, the tenants were degraded and their ancient rights taken away, and a bitter war was waged on Irish industry by the English

[1] C.S.P., Ire., 1603-6, p. 580.
[2] C.S.P., Ire., 1606-8, p. li.

Parliament. These ten years constitute the *Period of Redestruction*.

Thus, three times in the space of a hundred years, Irish economic life was shaken to its foundations, and the orderly and continued progress of the country towards an increased prosperity was rudely interrupted. The characteristic feature of the seventeenth century in Ireland was that economic development was periodically impeded by political cataclysms. After each disturbance the country settled down to make the best of its new conditions; fresh labour and enterprise combined to build up anew what had been destroyed, and did so with such rapidity and perseverance that many writers have remarked the power of recuperation after distress as one of the distinguishing features of Irish history; but every such period of reconstruction was cut short by a fresh calamity, which undid in a few years the work of many. This successive ebb and flow of prosperity in Ireland is the subject dealt with in the following pages.

CHAPTER I.

THE PERIOD OF CONSTRUCTION, 1603-1641.

(1) *Generally*.

THE period which we are about to consider in this chapter comprises the reign of James I. and the first fifteen years of the reign of Charles I. We have called this the period of construction, because it was characterized by the building up of an economic life in place of that which had been destroyed in the Elizabethan wars. The two great features which stand out pre-eminently in this process of construction are the settlement of Irish tenures, associated with the name of Davies, and the improvement of commerce and manufactures, associated with the name of Wentworth.

The settlement of the land will be discussed at length in a later section of this chapter, and it is not necessary to deal with it here. The principal measures adopted were the abolition of the Irish customs of tanistry and gavelkind; the taking of surrenders, and the regranting of the lands of some of the clans; and the plantation of certain districts with British occupiers. The result of these radical changes was that, in spite of a great deal of injustice suffered by individuals, the land system of the country was improved from the economic point of view; tenants were protected by security of tenure, due to the survival of local and national customs; and the art of husbandry was improved. The visible sign of this improvement was the general rise in the value of land,

and in the number of years' purchase at which it was sold.[1]

At the same time, efforts were made to increase the prosperity of Ireland in other directions. England had not as yet been roused to jealousy of the growth of Irish trade, partly because Ireland was still too undeveloped and too poor to appear as a serious competitor in foreign markets, and partly because the full import of the mercantile system had not yet come to be generally appreciated. On the contrary, it was altogether in the interests of the English King that Ireland should be prosperous, as it would thereby prove the source of a large revenue which would help to render the Crown less dependent on the English Parliament for the sinews of government. The policy of encouraging the trade of Ireland with this object in view was consistently followed by Wentworth, in whose letters we find it clearly outlined and frankly avowed.

"The reasons which moved me," he writes in 1633, "were the consideration of our commodious situation; the increase of trade and shipping in these parts of your dominions, which seem now only to want foreign commerce to make them a civil, rich, and contented people; and consequently more easily governed by your Majesty's ministers, and the more profitably for your Crown than in a savage and poor condition."[2] Four years later he wrote:—"I shall not neglect to preserve myself in good opinion with this people, in regard I become better able thereby to do my master's service; longer than it works to that purpose, I am very indifferent what they shall think or say concerning me. However, I cannot dissemble so far as not to profess, I wish extreme much prosperity to them also, and I should lay it up in my opinion as a mighty honour and happiness to become in some degree an instrument of it."[3] Wentworth was anxious that all unreasonable and vexatious restraints upon the

[1] Strafforde's *Letters*, vol. ii., p. 76; Brereton's *Travels in Ireland* (1635).
[2] Strafforde's *Letters*, vol. i., p. 93; vol. ii., pp. 19-20.
[3] *Ibid.*, vol. ii., pp. 121-2.

export of native commodities should be relaxed;[1] and was confident that, with proper encouragement, the trade of Ireland could be made a source of great benefit to the Crown. "That Kingdom is a growing people that would increase beyond all expectation if it were now a little favoured in this its first spring, and not discouraged by harder usage then either English or Scotch found."[2] When sending Charles an ingot of silver, as an outward and visible sign that the resources of Ireland were beginning to be worked, Wentworth gave expression to the hope that "this Kingdom now at length in these latter ages may not only fill up the greatness and dominion, but even the coffers and Exchequer of the Crown of England."[3] "I foresee this Kingdom is growing apace," he writes in 1637, "and a thousand pities it were by bringing new burthens upon them, to discourage those that daily come over and must make it flourish, especially when but by a short forbearance, till they have taken a good and sound root, his Majesty may at last gather five times as much from them without doing any harm, where a little pulled from them at first breaks off their fruit at its very bud."[4]

It was thus part of Wentworth's policy to develop the resources of Ireland in such a way as to render the country dependent on the Crown. "This is a ground I take with me, that to serve your Master completely well in Ireland, we must not only endeavour to enrich *them*, but make sure still to hold them dependent upon the Crown, and not able to subsist without *us*."[5] "I am of opinion that all wisdom advises to keep this Kingdom as much subordinate and dependent upon England as is possible, and holding them from the manufacture of wool, and thus enforcing them to fetch their clothing from thence, and to take all their salt from the King, how can they depart from us without nakedness and beggary?"[6]

[1] *Ibid.*, vol. i., pp. 308-380. [2] *Ibid.*, vol. ii., p. 20. [3] *Ibid.*, vol. i, p. 174.
[4] *Ibid.*, vol. ii., p. 89. [5] *Ibid.*, vol. i., p. 93. [6] *Ibid.*, vol. i., p. 193; vol. ii., p. 19.

Here, then, we have the expression of a clear-cut and definite policy towards Ireland. Irish prosperity is to be promoted so that a large revenue may be gathered for the English Crown, but no development must be permitted to take place which might tend to render Ireland politically independent of England. This policy of Wentworth compares favourably with that of later generations of English statesmen, in so far as it aimed at the improvement rather than the impoverishment of Ireland; but it should nevertheless evoke not the praise, but the condemnation of students of Irish history, because it was founded on the assumption that Ireland should be governed, not in its own interest, but in the interest of England.

The policy had the desired effect, and Ireland flourished during the reign of the first two Stuarts. In spite of the complete ruin of all the machinery of economic life at the beginning of the century, a marked improvement began to be noticed soon after the restoration of peace.[1] This rapid recuperation, which has been an essential feature of Irish history in every age, was rendered possible by the simplicity of the economic life of the country. In 1613, a petition of the nobility referred to "the present prosperity of Ireland."[2] "Ireland," we read in a proclamation dated 1627, "by reason of the peace and plenty it hath enjoyed and now doth enjoy, is stored and replenished with divers goods and profitable commodities and merchandizes."[3] In 1632 food was cheap, and the country "well stocked with corn and cattle."[4] "Their trade, their rents, their civility increase daily," wrote Wentworth two years later;[5] and in the following year land was known to be sold at twenty years' purchase.[6] Carte was of opinion that no such prosperity had ever before been attained as during the first forty years of the seventeenth century.[7] The following

[1] C.S.P., Ire., 1606-8, p. 2. [2] C.S.P., Carew. 1603-24, p. 265.
[3] Rymer's *Foedera*, 1743, vol. viii, pt. 2, p. 205; *Wood MSS.*, p. 192.
[4] C S.P., Ire., 1625-32, pp. 577, 674.
[5] Strafforde's *Letters*, vol. ii , pp. 80, 434-5; vol. i, p. 83.
[6] Brereton, *Travels in Ireland* (1635). [7] Carte, vol. i, p. 309.

sections deal in detail with special departments of the economic life of the country during this period.

(2) *The People and their Industrial Character.*

It is quite impossible to arrive at any conclusion as to the population of Ireland in 1600, but we know that the country must have been underpopulated owing to the terrible wars which it had endured, and the great destruction of life which had taken place by famine. The population was further weakened in the first half of the seventeenth century by emigration. The famous flight of the Earls was followed by the departure of a great many gentlemen and their retainers; but, apart from exceptional incidents of this kind, there was a continuous stream of Irish emigrants to England and the Continent as a result of the loss of their land which many had suffered, and the lack of employment which no doubt existed in consequence of the shattered condition of the country. Numbers of Irish beggars were to be found in France,[1] and in 1606 London and the surrounding country was pestered with Irish beggars.[2]

It is impossible to say at what rate the population increased in the years 1600-41, and all we know of the population at the end of that period is that it was insufficient for the economic needs of the country. In 1641 it was proposed to send sixteen thousand men abroad for the King of Spain's service, but the Irish Lords and Commons objected to this, as "there were not men enough in the country to maintain agriculture and manufacture."[3] A rough guess may be made at the population in 1641, if Petty is considered reliable when he says that the population in that year was greater than in 1687, when he estimated it at one million three hundred thousand.[4]

[1] C.S.P., Ire., 1606-8, p. 98; *Corporation of Rye MSS.*, p. 134.
[2] C.S.P., Ire, 1603-6., pp. 462, 487.   [3] *I.C.J.*, vol. i., p. 276.
[4] Petty, *Treatise of Ireland.*

IN THE SEVENTEENTH CENTURY. 13

The Catholics were not at this time subject to any of the disabilities which affected them economically, either by preventing them from acquiring wealth, or by robbing them of the ambition which is essential to maintain the industrious character of a people. The few penal laws which existed were directed simply and solely against the Catholic religion itself, and rested on a purely political foundation; they were not in any sense of the word laws designed, as were the famous penal laws of the eighteenth century, "to make the Catholics poor and to keep them poor." The only laws dealing with religion on the Statute Book at the time were the laws making the King supreme head of the Church;[1] prohibiting appeals to Rome;[2] or the recognition of the Pope's authority;[3] and the Act of Uniformity prescribing the use of the book of Common Prayer and the obligation to attend the Church of Ireland services.[4] The only occupations which were closed to Catholics on account of their religion were those of professors in universities, schoolmasters, and private tutors, and certain public employments where it was feared that obedience to the Pope's authority was incompatible with due obedience to the Crown. Strictly speaking, a lawyer could not take a degree in law or plead at the bar without taking the Oath of Supremacy, but Catholics were permitted to carry on a chamber practice at which they could make a substantial livelihood, and, as a matter of fact, these laws were enforced with the utmost laxity, and were frequently more honoured in the breach than in the observance.[5] Apart from these laws, there were no penal laws in the modern sense at all, and the Catholics of Ireland at that date were under fewer disabilities than the Catholics of England.[6] Even in the corporate towns, which, at a later date, became the strongholds of Protestantism, the Catholics peaceably pursued their trades and rose to positions of importance.[7]

[1] 28 Henry VIII., c. 5.   [2] 28 Henry VIII., c. 6.   [3] 28 Henry VIII., c. 13.
[4] 2 Eliz., c. 2.   [5] Carte, vol. i, pp. 86, 88 and 309.
[6] Carte, vol. i., p. 68.   [7] Gale, *Corporate System in Ireland*, p. 50.

The evil of begging does not seem to have been very marked at this time, probably owing to the sparseness of the population and the consequent absence of unemployment. In 1631, houses of correction were set up, in which youths found begging were taught trades;[1] but begging was an evil complained of but little at this time as compared with the eighteenth century.[2] The only class of idlers in the community which called for attention were the dispossessed gentry, who were accustomed to travel around the country without any occupation, and to stay with their retainers and followers in different houses, against which practice a statute was directed in 1634.[3] The important thing to notice is that in the seventeenth century no persistent accusations of laziness or idleness were made against the Irish people; all the evidence we possess is to the contrary. Sir John Davies was of opinion that any idleness which existed in Ireland was caused by the insecurity of life during the wars.[4] "The natives are apt to labour when they may have hire and reward for the same, but the Irish lords and gentry do never give the poor people anything for their labour, which they think shall dispose them to idleness."[5] We also have Gookin's testimony that the poor Irish were very skilful labourers, and apt in overcoming difficulties.[6]

The industrious character of the people in this period is probably accounted for by the fact that the tenants were secured in their holdings by tenant-right, and that there were no penal laws to sap the industrious spirit of the Catholics; and there was consequently no discouragement to work such as existed in the eighteenth century, when the Irish people generally came to be regarded as a lazy people. It is also possible that the introduction of potatoes had something to do with this change of character, as it enabled the cottier class to obtain a subsistence with a minimum of labour.

[1] C.S.P., Ire., 1625-32, p. 611.   [2] C.S.P., Ire., 1603-6, p. 135.   [3] 10 & 11 Car. I., c. 15.
[4] Davies' *Discoverie*, pp. 669-710.   [5] C.S.P., Ire., 1647-60, p. 129.
[6] *The Great Case of Transplantation Discussed*, London, 1655.

But if the Irish were not idle, they suffered from another vice which tended to impair their economic efficiency. "Drunkenness is the only curse of the country," we read in 1600;[1] and at a later date the drinking of usquebaugh and aqua vitae was said to be extraordinary.[2] "The monopolies of selling wine by retail and aqua vitae must amount to a very great sum, by reason the Irish, and chiefly the mere Irish, give themselves unreasonably to both kinds, especially the latter liquor."[3] An Act was passed in 1634 attempting to regulate the excessive number of ale houses, wherein it was recited that the country was full of houses where drink could be obtained, and that excessive drinking was its national misfortune.[4] The Act, however, was not strictly observed, for we read some years later that there were still many ale houses in the woods and bogs which paid no duty.[5]

(3) *The Settlement of the Land System.*

In a consideration of the changes which took place in the ownership and tenure of land in Ireland at the end of the sixteenth and beginning of the seventeenth century, the features which loom largest in the popular mind are the confiscations and plantations. Of course, these occupied a very large place in the Irish history of that period, and are undoubtedly of the greatest political importance, as it was the change of ownership brought about by these confiscations and plantations which changed the character of the land system of Ireland, and laid the foundation of the agrarian disturbances of the succeeding centuries. But, although these changes were of great importance, they were possibly not so important, from the economic point of view, as the changes which took place about the same time in the relationship between the chiefs and the clans, whereby the members

[1] C.S.P., Ire., 1600-1, p. 456. [2] C.S.P., Ire., 1647-60, p. 169.
[3] *Advertisements for Ireland*, MS., T.C.D. [4] 10 & 11 Car. I., c. 5.
[5] C.S.P., Ire., 1647-1660, p. 342; Strafforde's *Letters*, vol. i., p. 192.

of the ancient Irish land-owning class were either elevated to the position of freeholders, or degraded to the position of tenants, instead of being a middle-class as under the Brehon system. As we shall see, there was also a large section of the population dependent on the land for a livelihood and intimately connected with it who were not themselves landowners; these formed the substratum of Irish life, the ancient churls of the Irish tenure, who continued to subsist at the bottom of the social scale as the day labourers and cottiers of later centuries.

The change in the relationship between the chiefs and the clans was carried out partly during the reign of Elizabeth, and partly during the reign of James I., by means of the policy which came to be known as "surrender and regrant." This policy was to induce the chief or the clan to make a surrender of lands to the Crown and then to make a regrant in one of two alternate forms; one was a regrant of the whole territory inhabited by a particular clan to the chief of the clan; the other a regrant of the clan lands to the chief and members of the clan, as far as possible in proportion to their respective rights within the clan. There were also in Elizabeth's reign some efforts to establish plantations in Munster, but in their ultimate results these plantations were not of very much importance. The latest and best authority on the subject is of opinion that only 200,000 acres were given to the undertakers in Munster, and that the project of a great English colonization was defeated. Besides, as we shall see, the undertakers did not fulfil the conditions laid down for them; far fewer English families were brought over than had been arranged for; and Irish tenants were brought in to fill the gaps.[1] It will, therefore, be best to consider the change in the Irish land system without complicating the matter by any further reference to the plantation of Munster, and to devote our attention to the more important part of the subject, namely, the policy of surrender and regrant.

[1] Butler, *Confiscations in Irish History*, pp. 29-30.

From the English point of view there were many arguments in favour of each of the two principles upon which the policy of regrant was carried out. The grant of the whole clan-land to the chief was favoured because it would induce the chief to surrender, and would bind him to fidelity to the Crown. There was always, moreover, a possibility that the chief would break away from his allegiance, and thereby give the Crown a legal excuse for forfeiting the whole revenues of the clan-land. On the other hand, the argument for regranting the lands to the clans in proportion to the clansmen's former rights, *inter se*, was that the individual chiefs would not become too powerful, and that it would, by creating numerous freeholders, bring the clans directly under the dominion of the Crown. This was the policy which dictated the statute of Quia Emptores in England several centuries before, and it was an avowed principle of English policy in Ireland at this time:—" The poor Irish," wrote St. John to Salisbury, in 1607, " estimate more thair landlord that they know than their King whom they seldom hear of; when they shall be inured to know that they hold their lands immediately of the king they will neglect their chiefs, whom only they love now, and only turn their affection and loyalty to the King.'"[1] " It hath been the constant endeavour of this State to break the dependences which great lords draw to themselves of followers, tenants, and neighbours, to make the subject to hold immediately of the Crown.'"[2] This policy, however, was only carried out to a certain point. Only the chief and the more important clansmen were reinstated as freeholders, the lesser members of the clan being, as a rule, passed over in the settlement, probably because of the objection of statesmen of the time to the creation of anything like a peasant proprietorship. " The multitude of small freeholders beggars the country," was a statement generally believed at this time.[3]

[1] C.S.P., Ire., 1606-8, p. 304. [2] Strafforde's *Letters*, vol. ii., p. 366.
[3] Butler, *The Policy of Surrender and Regrant*, Jl. R.S.A.I., vol. xliii, p. 101.

The regrant of lands to the chiefs absolutely is represented by some writers on the history of the period as in all cases unreasonable, and as a premeditated system of robbery on the part of the English Government, but a more thorough knowledge of the conditions of Irish land-holding in the sixteenth century convinces us that this is taking an unnecessarily prejudiced view. As a matter of fact, the chiefs had in many instances themselves become the absolute owners of the lands belonging to their clans, partly, thanks to certain principles of the Brehon law, and partly, thanks to the strong hand. Many of the chiefs appropriated large parts of the clan lands for the aggrandisement of themselves and their families—in some cases violently and suddenly, in other cases gradually. Again, when the clans were driven out from their original homes to settle in new territories, it would seem that, by Irish law, lands thus acquired under the leadership of the chief became the property of that chief to distribute as he pleased. This principle was held to apply not only in the case of clans moving to new districts, as in the case of the O'Byrnes and the O'Tooles, but also in the case of a clan reconquering a district from which it had been expelled by the English, as in the case of the O'Ferralls of Longford, the O'Kellys of Galway, and the O'Dowds of Sligo. In this case the old rights of the clan were held to be extinguished by the English conquest.[1] It was, therefore, sometimes a matter of difficulty for the English authorities to decide whether the land surrendered should be regranted to the chief alone or distributed among the clan. We find Sir John Davies (who, however, did not consider abstract justice alone) expressing the difficulty he found in deciding such a question:—" But touching the inferior gentlemen and inhabitants, it was not certainly known to the State here whether they were only tenants-at-will to the chief lords, whereof the uncer-

---

[1] Butler, *Policy of Surrender and Regrant*, pp. 103-6.

tain cutting which the lords used upon them might be an argument, or whether they were freeholders yielding of right to their chief lord certain rights and services, as many of them do allege, affirming that the Irish cutting was an usurpation and a wrong. This was a point wherein the Lord Deputy and Council did much desire to be resolved."[1] Of course, there were also other motives besides the mere desire to do justice which operated to procure grants of the whole clan territory for some of the chiefs. " The special services of particular chiefs, the caprice of Lord Deputies, were some of the factors which explain why some chiefs received grants of the entire clan territories."[2]

Examples of both kinds of regrant may be found in the records of the reign of Queen Elizabeth. Some of the clan lands were granted to the chief absolutely, without any provision for the inferior members of the clan. Examples of these are the grants to MacCarthy of Muskerry, O'Neill and O'Donnell. " It so happened that the chiefs who thus filched the lands from the people were those who stood most prominent in the public eye. Hence the idea, repeated in book after book on Irish history, that the clansmen were robbed of their lands by Elizabeth, to satisfy the greed of the chiefs."[3] But in by far the greater number of instances in Elizabeth's reign, the other policy was the one adopted. In 1591 MacMahon's country was regranted amongst the clansmen, the leading men getting large estates with chief rents from the lesser proprietors, who were themselves confirmed in the lands which they already held by Irish custom. All, great and small, were to hold direct from the Crown.[4] A similar settlement was attempted on the lands of Sir Arthur Magennis of Iveagh. Of course, the most famous example was the Composition of Connacht and Clare in 1685, which was so successful in

---

[1] Davies, *Letter to the Earl of Salisbury*, 1607.
[2] Butler, *The Policy of Surrender and Regrant*, p. 106.
[3] Butler, *The Policy of Surrender and Regrant*, p. 102.
[4] C.S.P., Ire., 1591, p. 428.

creating a large number of medium-sized freeholders that it is generally supposed to have been the reason why Connacht and Clare were the most peaceable and least rebellious parts of Ireland during the outbreaks of the seventeenth century.[1] There was no uniformity, however, in the policy adopted, and the greatest varieties of policy were to be found in neighbouring baronies. For instance, in Duhallow the chiefs got all or most of the clan lands; in Carbery the rights of the clansmen were completely respected; while Desmond was settled on a principle intermediate between the two.[2] However, the one principle which pervaded all the settlements—although not quite so noticeably in Connacht and Clare—was the degradation of the lower grades of the clan from the position of clansmen to that of tenants.

"The net result arrived at during the reign of Elizabeth was that the main lines for the settlement of the land had been laid down for a great part of the island. Some of the more influential lords had obtained all the clan lands; over a large part of Ireland it had been decided that they were to be satisfied with the demesne lands set aside by the clan to provide for the maintenance of the chief. In some cases the grants were so vaguely worded that it was quite uncertain what had been granted. Everywhere the constant warfare which went on during Elizabeth's reign interfered to prevent a thorough settlement." Here, as always, we find the orderly economic development of Ireland turned aside and checked by the pressure of political events.

The outcome of Elizabeth's reign then was that the bulk of Irish land remained in the hands of the Irish or of the old Anglo-Irish. Although the relationship *inter se* of the Irish land owners had been to a large extent altered, there had been no systematic attempt, with the commonly exaggerated exception of Munster, to replace the Irish landowners by English settlers. More-

[1] Sigerson, *History of Irish Land Tenures*, p. 31.
[2] Butler, *Policy of Surrender and Regrant*, p. 113.

over, the legal changes in ownership had not succeeded in disturbing the customary rights which prevailed throughout the country, and the relationship of the inferior and superior owners was universally regulated by the old Irish customs down to 1600.[1]

Under James I. there was a sinister change of policy which took the form of systematic attempts to plant the country with English tenants, and to degrade the former owner to an inferior position.

In the new scheme all respect for justice—a respect which, on the whole, had marked the Tudor dealings with the land—was thrown aside. " Henry VIII. had abandoned the old plan of forcible dispossession of the natives; he had laid down the principle that the Irish were to be given a legal title to the lands they actually occupied. Sir John Davies, writing in 1607, had declared that the State had never taken hold of a title derived from conquest against such of the Irish as had not been deprived of their lands at the first conquest, but were permitted to die seised of the same in the King's allegiance. This is true of Tudor days, and of the early years of James. The Tudors had encouraged the lords of Irish countries to make surrenders of their lands with a view to getting a legal title to them. But no force was used to compel them to do so, and, as is clearly shown in the cases of Donegal and Carbery, the Crown did not disturb in their possessions either the chiefs or the clansmen of those territories where no such surrender had been made.

" Now all was changed. Old grants dating from the time of the first invasion were raked up to show that the Crown was entitled, either as heir of the Mortimers, or under the Statute of Absentees, or through the treason and forfeiture of nominal English owners, to the greater part of the territory in Leinster still inhabited by the natives."[2]

A very important step in the policy of Sir John

---
[1] Sigerson, *op. cit.*, pp. 24-5; Butler, *Confiscations in Irish History*, pp. 35-6.
[2] Butler, *The Policy of Surrender and Regrant*, p. 122.

Davies was the abolition of the Irish customs of tanistry and gavelkind, which was effected by two judicial decisions in 1605. It is possible, indeed almost certain, that the ultimate effects of these decisions were not apparent to those who pronounced them, and it is probably unfair to say that the Judges were induced to decide as they did in order to clear away the difficulty which stood in the way of Sir John Davies' Ulster policy.[1] As a matter of fact, the decisions were made before the opportunity for a plantation of Ulster had presented itself, and there was, therefore, probably no connection between the two events. The abolition of gavelkind and tanistry, however, undoubtedly proved a very useful weapon in the hands of Davies, when, at a later date, he contrived the great scheme of the plantation of Ulster. At first it was his intention to settle Ulster on Elizabethan lines, but at a later date he conceived the brilliant idea that, owing to the recent decisions on gavelkind, the Irish customs of inheritance could not be reduced to agreement with the Common Law of England, and that, therefore, the natives of the confiscated counties were only tenants-at-will of the lords, and consequently were possessed of no rights as against the Crown.[2] The cynical injustice of this theory is illustrated by the fact that, while the jurymen, whom Davies had himself empanelled to declare the Earls who had fled from Ireland traitors, were considered by him as freeholders, it was these very freeholders whom he now elected to treat as mere squatters. It is, therefore, correct to date the first beginnings of the Stuart policy of confiscation and plantation, as distinguished from the Elizabethan policy of surrender and regrant, from the decisions on gavelkind and tanistry, the effect of which was "to reduce the Irish claimants of land in virtue of the two customs condemned to the position of mere squatters."[3]

---

[1] This suggestion is made by Dr. Sigerson in his *History of Irish Land Tenures*.
[2] Butler, *The Policy of Surrender and Regrant*, p. 121; *Confiscations in Irish History*, p. 45.   [3] Butler, *Confiscations in Irish History*, p. 58.

It is no part of the plan of this book to describe in detail the various plantations which were carried out in the reign of James I.; that has been repeatedly done before, and the only object of the present writer is to examine the effect which these plantations had on the subsequent economic development of the country. It is important, however, to remember that the treatment accorded to the Irish natives was different in the different plantations; and also that the treatment which they, in fact, received, was in many cases harsher than that intended by those who devised the plantations.

With regard to the Ulster plantations, the different counties planted were treated differently in this respect. In Fermanagh 63 natives received grants of land in freehold, Conor Roe Maguire alone receiving a grant of 7,000 acres of profitable land. In Cavan, on the other hand, only 39 natives received grants, the rest of the population losing all their property, and being driven either to emigrate or to seek a livelihood as tenants on the lands granted to servitors, natives, and the Church, who were the only classes permitted by the new grantees to lease lands to Irish tenants.[1] In the other four counties the rights of the natives were respected to an even smaller degree; only 153 Irish received grants in Armagh, Tyrone, and Donegal; whereas very few, indeed, were granted any land in Londonderry.[2] According to one list, of the 511,000 acres at which the six counties were estimated, the natives got only 52,000 acres, or slightly more than one-tenth.[3]

In the Wexford plantation, 667 natives claimed, but only the claims of 440 were admitted. However, it was one thing to have one's claim admitted, and another to receive a grant. Of the 440 whose claims were admitted to be good, only 57 received grants at first, but on a subsequent inquiry the frauds committed against the natives were admitted by the Government to be so gross

[1] Butler, *The Policy of Surrender and Regrant*, p. 121.
[2] Butler, *The Policy of Surrender and Regrant*, pp. 121-2.
[3] Butler, *Confiscations in Irish History*, p. 46.

that 80 more grants were made. All the other natives were deprived of their lands on the ground that they claimed less than 60 acres, and that the creation of small freeholders was undesirable. The solution of the problem presented to these dispossessed landowners of finding a livelihood was solved by a benevolent Government; on their going to Dublin to complain of the treatment they had received, they were shipped off as slaves to Virginia.[1]

In the plantation of Longford, one-fourth of the land was allocated to planters, the other three-fourths remaining to the natives, but no native was to be allowed to claim for less than 60 acres. The result was that only 142 natives received land, and that many of the remainder, " after the loss of all their possessions there, ran mad, and others died instantly of grief, and others who, on their death-beds, were in such a taking that they, by their earnest persuasions, caused some of their family and friends to bring them out to have a last sight of the hills and fields they lost."[2]

The Leitrim plantation was carried out on the same lines, the smaller freeholders being dispossessed, and half of the lands being assigned to new-comers. Thus, all the small landowners lost everything.[3] There were other smaller plantations made in various parts of Ireland in which the same features are presented. There was one sinister feature, however, about these later plantations which savours of the degrading policy which became so prominent later in the century. The native grantees were usually forbidden to sell or alienate their lands to any Irish purchaser, or to grant Irish tenants longer leases than for three lives or forty-one years.[4]

Of course, it must be remembered that the theoretical rights of the native Irish under these plantations were not always respected in fact, and that great frauds were

[1] Butler, *Policy of Surrender and Regrant*, p. 123.
[2] Butler. *Policy of Surrender and Regrant*, p. 123.
[3] Butler, *Confiscations in Irish History*, pp. 86-87.
[4] Butler, *Policy of Surrender and Regrant*, pp. 123-4; *Confiscations in Irish History*, p. 90.

committed against them by the English Commissioners who had immediate charge of the settlement. We have seen that the frauds in Wexford were notorious; and in Ulster the Irish did not, in fact, receive even the small proportions which were by law to be assigned to them.[1]

We have now very briefly reviewed the main points of the Elizabethan and Jacobean policy of land settlement in Ireland. There is no necessity to refer to the threatened confiscation of Connacht, which loomed so largely in the political history of the time, because, as is well known, circumstances occurred which prevented its being carried out. It, therefore, had no immediate economic significance in itself, but, of course, its historical significance was stupendous, inasmuch as it was the direct occasion of the rising of 1641, which resulted in the complete destruction of the economic life of the country.

The net results of the juggling with lands, which had gone on for over a hundred years, were (1) that the Irish small freeholders had been crushed out of existence; (2) that there was an absolute diminution in the quantity of land owned by the Irish, but, in spite of the many plantations, they still owned the greater part of the country in 1641; (3) that Irish tenants had acquired rights in the lands even in the plantation counties; and (4) that these tenants were protected in their tenant right by the old Irish customs which had survived even on confiscated lands. Some elaboration of each of these points is necessary.

We have seen that part of the plantation scheme was the crushing out of the small freeholder. In Longford, Wexford and Leitrim no clansmen who presented a claim to less than 60 acres was admitted, and it was these people and their descendants who subsequently emigrated or sank to the condition of tenants, or, worse still, disdaining to live as the tenants of strangers and of their former equals in the clan, became wood kerns, rapparrees

[1] C.S.P., Ire., 1647-60, p. 204; *Advertisements for Ireland*, MS., T.C.D., 1622.

and tories—the precursors of Captain Rock. In one settlement alone were the rights of the small freeholders to some extent respected, namely, in Connacht, and there is no doubt that the peaceable attitude of the Connacht peasantry in the Rebellion was the result of this establishment of what was almost a peasant proprietary.[1]

The fact that there was an absolute diminution in the quantity of land owned by the Irish calls for no demonstration, as large parts of the territory of the country had been planted with British freeholders; but in spite of the many plantations, the greater part of the country still remained in Irish hands in 1641. According to the abstract compiled by Mr. W. H. Hardinge, published in the *Transactions of the Royal Irish Academy*, out of the 20,000,000 statute acres in Ireland, 11,000,000 acres belonged to Catholics in 1641.[2] " From the list printed in O'Hart's *Irish Landed Gentry when Cromwell came to Ireland*, of ' Forfeiting Proprietors Listed,' made in 1657 by Christopher Gough, we find that in fifteen counties and one barony there were some 4,120 landowners whose estates were confiscated. Now, as these included five of the six Ulster plantation counties, in which there were only about 270 Irish Catholic landowners in 1641, and as nearly all Connacht and four Munster counties were omitted, we may safely assume that in the seventeen counties for which we have no return, there were at least 6,000 landowners Catholics—*i.e.*, of Irish or Anglo-Norman descent—in 1641."[3]

Irish tenants crept in, in spite of all the elaborate provisions which were made to keep them out. The original intentions with regard to the Munster plantation were found at an early date to be quite impossible of fulfilment. As early as 1589 we find a complaint that the Undertakers found " such profit from their Irish

---

[1] Sigerson, *op. cit.*, p. 31.
[2] *Trans. R.I.A.*, vol. xxiv., ant. p. 7.
[3] Butler, *Policy of Surrender and Regrant*, p. 126.

tenants who give them the fourth sheaf of all their corn and 16d. yearly for their beasts' grains, so they care not though they never plant an Englishman there."[1] Spenser complained that, "instead of keeping out the Irish, the Undertakers did not only make the Irish their tenants in these lands, but thrust out the English."[2] In County Limerick, in 1600, the "Weeds," that is, the Irish, were said to have cast the English out;[3] and ten years later the majority of the tenants on the lands planted in Ireland were Irish.[4]

In Ulster the same thing occurred:—" The new landowners could not cultivate their demesne lands without Irish labour; Irish tenants offered higher rents than could be obtained from British; these last could not always be obtained. At first extensions were obtained of the time within which the Irish occupiers were to remove. Then the new owners, by a policy of passive resistance, succeeded ultimately, in spite of numerous efforts on the part of the Government, in evading this, one of the fundamental conditions of their grants. Finally, in 1626, permission was given to take Irish tenants on a quarter of the Undertakers' lands, provided that they were given leases for life, or for twenty-one years, and that provision was made to force such Irish tenants to abandon the mode of life and the religion of their forefathers."[5] As early as 1608 complaints were made of the large rents given by the Irish, who were outbidding the English.[6] In Pynnar's account of the planted lands prepared in 1618, we read of "an infinite multitude of Irish on the land in Ulster who give such high rents that no English can get land."[7] The following list[8] shows the number of English and Irish tenants on the lands belonging to the London companies, which were, above all others, meant to contain a thoroughly English settlement to the

[1] Payne, *Description of Munster in* 1589, Irish Arch. Society, vol. i.
[2] Spencer, *View of the State of Ireland.*
[3] C.S.P., Ire., 1600-1., p. 137.
[4] C.S.P., Ire, 1611-14, p. 219.
[5] Butler, *Confiscations in Irish History,* p. 47.
[6] C.S.P., Ire., 1606-8, p. 304; and 1611-14, p. 254.
[7] Harris, *Hibernica*, pp. 76, 86, 93, 120.   [8] C.S.P., Ire., 1615-25., p. 471.

exclusion of the Irish; and shows to what extent this intention was frustrated :—

	Townlands.	Planted with Irish Tenants.	Planted with English Tenants.
1. Salters	53½	42½	11
2. Vintners	49½	29	20¼
3. Drapers	64	48	16
4. Mercers	47	29	18
5. Goldsmiths	42¾	17	24¾
6. Grocers	53	21	32
7. Fishmongers	55	24	31
8. Haberdashers	57½	17½	40
9. Clothworkers	48½	8½	40
10. Merchant Taylors	47	24	23
11. Ironmongers	47	30½	16½
12. Skinners	43	14	29

The same feature of the crowding in of Irish tenants also prevailed on the plantations of Leinster.[1] "The Undertakers all for the most part especially of our richer sort and the Corporations here of England and others here resident that have land there suffer the under sort of mere Irish to inhabit their lands because they pay them greater rents they say than the British will; so as those Undertakers being the eminenter men and the abler giving such a precedent to the rest many by their imitation begin to do the same."[2]

The presence of so large a number of Irish tenants, who were doubtless imbued with a knowledge of and fondness for the old customary system which they remembered so well, prevented the extinction of the Irish customs. There is evidence that, in spite of the decisions of the courts, the old custom of gavelkind still continued to be practised in the country. The Books of Survey and Distribution, printed in O'Hart's *Irish Landed Gentry when Cromwell came to Ireland*, plainly show the existence of collective ownership in Connacht in 1641.[3] More important than the survival of the custom of gavel-

---
[1] Butler, *Confiscations in Irish History*, p. 90.
[2] *Advertisements for Ireland*, MS., T.C.D., 1622.
[3] Butler, *The Policy of Surrender and Regrant*, p. 110.

kind was the recognition of the principle of fixity of tenure, which prevailed throughout the country, so that it is probably true to say that in the seventeenth century what later came to be known as the Ulster custom extended all over Ireland.[1] Apparently these customary rights grew out of the principle of the Brehon laws by which possession of land for a certain time gave the occupier a right to continue his tenure thereof.[2] The security which this custom was felt to give is evidenced by the dislike which Irish occupiers of land evinced to the taking of a definite lease, and by their preference to rely on the custom. It was, on the other hand, part of the English policy to make the Irish tenants take leases for a long period, probably with the idea of binding them down to the land, and thus abolishing the system of " creaghting," or driving herds of cattle from place to place. In 1610 it was suggested that any lease in Ireland for less than ten years should be void;[3] in the plantation of Leitrim instructions were given that those natives who were not made freeholders should be given leases for forty years.[4] But the Irish persistently refused, if possible, to take leases, and preferred to be treated as tenants from year to year, a conclusive proof of the security which they felt in the " custom of the country " —a custom which was enforced, no doubt, by the predecessors of Captain Rock.[5]

From the economic point of view, therefore, nothing had been done before 1640 to render Irish land tenures unbearable to the tenants. We are not concerned with the justice or injustice of the plantations and confiscations. Morally, of course, they were absolutely unjustifiable, as they amounted to sheer robbery, but the economist must regard them as mere changes in possession, and the only question with which he is concerned

[1] Sigerson, p. 65.
[2] Hore, *The Archæology of Irish Tenant Right*, Ulster Jl. of Arch, o.s. vol. vi., p. 109. [3] C.S.P., Carew, 1603-24, p. 162.
[4] *Desiderata Curiosa Hibernica*, vol. ii., p. 52.
[5] Butler, *Confiscations in Irish History*, p. 90; Hore, *Archæology of Irish Tenant Right*, Ulster Jl. of Arch, o.s. vol. vi., p. 109. In Ulster the Irish tenants took leases from year to year, *Pynnar's Survey*, Harris, Hibernica, p. 77.

is whether the new tenant cultivated the land with greater success than the old. Economically, it is more important that land should be rightly used by a wrongful owner than wasted or abused by the rightful owner. From this point of view, indeed, the substitution of English tenures was probably a good thing, as it tied the landowners down to a particular portion of land which was not liable to be changed at certain or uncertain intervals, and tended to abolish the Irish system of creaghting, which was economically so unsatisfactory. It is possible that if the rights of the small freeholders had been universally acknowledged, as in Connacht, and if the revolution in landownership, which was designed as a measure of reconstruction, had not to a large extent been used as a weapon for exploitation, a peasant proprietary might have been established at an earlier date and at much less cost than it has been. It is also possible that the English Undertakers were more advanced in the art of husbandry than the Irish whom they had supplanted,[1] but, even as it was, a large number of freeholders were established, and it could not be said that the lands had become engrossed by a few large proprietors.

This must not be taken as casting any discredit on the Irish system of agriculture: as we have seen in the Introduction, Ireland was, in early times, a fruitful and prosperous land, but it is very probable that the skill of earlier times had been lost in the perpetual warfare of the sixteenth century. There was, therefore, a possibility that the land system which was coming into being in the seventeenth century might become stable, especially when it is remembered that tenant-right was very general, and that the peculiar evils of the later Irish land system, which flowed from an absence of tenant-right, had not developed. Above all, there was no religious disability to hold land, and, therefore, none of the systematic degradation and inevitable low state of

---

[1] This would appear to have been so from *Pynnar's Survey*.

agriculture which were the worst features of Irish life in the eighteenth century. " It being most sure that the lower sort of the Irish subject hath not in any age lived so preserved from the pressure and oppression of the great ones, as now they do, for which they bless God and the King, and begin to discern and taste the great and manifold benefits they gather from their immediate dependence upon the Crown, in comparison of the scant and narrow coverings they formerly borrowed from their petty yet imperious lords."[1]

Economically, therefore, it is probably true that the land system at this time was an improvement on the system of the previous century,[2] and that it would have been a success had Ireland been a country whose progress depended on the evolution of economic laws alone. But, unfortunately, Ireland at that date was not composed of a community of " economic men "; and a system which might have proved beneficial if such a community had existed, contained within itself the germs of its own dissolution when applied to a country involved in a racial war. The dispossessed land-owner was not content to watch from without the economic development of the country from the benefits of which he personally was excluded; on the contrary, he actively interfered with the country's progress. Whether he was called a kern, a rapparee, a tory, or a Captain Rock, he was all the time the same person—a person ever conscious of the wrongs which he and his ancestors had suffered in the confiscation of his lands, and ever anxious to strike a blow for the righting of that injustice, even if such righting entailed further wrongs and possibly greater suffering. In short, he was the person who rose in arms in 1641. It is, therefore, impossible to be satisfied with a system which, of its very nature, produced results fatal not only to the progress but the orderly existence of the country. " Owing to the political unrest of the period," says Dr. Cunningham, " there was little security for property, so

[1] Strafforde's *Letters*, vol. ii., p. 93. [2] See Carte, vol. i., p. 90.

that the first essential for progress was entirely wanting."[1]

The question of to what degree the rents of Irish land were remitted to England is one which cannot be left out of consideration in discussing the Irish land system at any period. Absenteeism had always been a grievance in Ireland from early times. In the reign of Richard II. an Act was passed forfeiting two-thirds of the profits of the estates of absentee landowners, and in 1537 all the lands of the houses of Norfolk, Shrewsbury and Berkeley, were resumed, as their proprietors had become perpetual absentees.[2] The objection to absenteeism in earlier times, however, was political, not economic; it was feared that if the great lords who had been granted Irish lands were absent from them, the lands would again fall into the hands of the Irish. Absenteeism as an economic evil had not developed at this time to any large extent, although some of its characteristic features had begun to be complained of in 1622. "That which hindreth most the success of the plantations," we read in a manuscript written in that year, "is that some which reside altogether in England do sometimes bestow their proportions upon their footmen and other of their meaner servants, and these are seated now and then in the principal seats and dwellings of the prime natives, and do overtop them with more sway and authority than their lord and master would were he there in person, and this discourages much the natives to industry."[3] In 1628 one of the Graces demanded of Charles I. was "that as the evils of absenteeism cause a great economic drain from Ireland, care be taken that the great land-owners be compelled to remain at least half the year in Ireland"; and the King's answer was that all the nobility and Undertakers should reside there unless summoned to England by the King.[4] In 1634 an Act was passed providing that all the nobility

[1] *Growth of English Industry and Commerce*, vol. ii., p. 362.
[2] Davies' *Discoverie*, pp. 614 and 688.
[3] *Advertisements for Ireland*, MS., T.C.D.
[4] C.S.P., Ire., 1625-32, p. 336.

of Ireland resident abroad should pay the same taxes as if present in Ireland.[1] On the whole, absenteeism in this period did not attain to any serious proportion; and, indeed, it was one of the evils specially aimed against in the schemes of the Tudor and early Stuart settlements. The area of English rule in Ireland had diminished in the fifteenth century largely on account of the continuous absence in England of the grantees of Irish land, whose business it was to subdue the country in consideration of their grants, and it was against a repetition of this danger that the schemes of Elizabeth in Munster and of Davies in Ulster and Leinster were particularly directed. Absenteeism again became an evil after the Restoration, but in the period under review in the present chapter, it was not very seriously or extensively complained of.

(4) *The Cottiers.*

The central economic problem of Irish life in the eighteenth and nineteenth centuries was the misery of the lower classes of the population. Books could be filled with an account of the misery which the cottiers suffered from cold, hunger and landlessness, and it would be quite impossible to exaggerate the wretchedness of their condition or the importance of the problem to which it gave rise. There is, however, one popular fallacy with regard to the evil of the cottier system. It is commonly thought and said that these wretched inhabitants of mud cabins had, in addition to their other sorrows, the bitter memory of a proud ancestry, and that they were the descendants of the ancient noble families of the country. This is only to some extent true. Of course, during the eighteenth century a certain number of farmers sank to the condition of cottiers, owing, in the first place, to the penal laws, which shut out Catholics from the possibility of being prosperous or skilful farmers, and secondly to the widespread conversion by the

[1] 10 Chas. I., Sess. 3, Chap. 21.

large land-owners of arable land into pasture which dispossessed so many tenants. But, generally speaking, the cottiers of the eighteenth century were the descendants of the cottiers of the seventeenth; the old Irish free clansmen for the most part emigrated when their lands were taken from them, and did not show any disposition to work as small farmers or labourers on the lands which they had formerly owned, but which now were theirs no more. When they remained in Ireland they drifted more naturally into the large class of idlers whom we read of in 1634 as "walking up and down in the country with one or more greyhounds, coshering or lodging or cessing themselves, their followers and greyhounds upon the inhabitants,"[1] or else they reinforced the ranks of Captain Rock's numerous army.

What is commonly forgotten is that there was always in Ireland a low class of agricultural labourers, who were the menials of the owners of the land, whether the clan system or the English system prevailed, and who formed what Dr. Sigerson[2] has aptly called "the settled substratum of Irish society." The old Irish civilisation rested essentially for its existence, as did the civilisation of Greece, on the existence of an unfree class, who had no rights in the land of the country—which is equivalent to saying that they had no property rights whatever.[3] These "churls," as they were called in English, were of much account to the planters of the sixteenth century:—
"There is no doubt but there will great numbers of the husbandmen which they call churls come to live under us and offer to ferme our grounds, for the churl of Ireland is a very simple and toilsome man, desiring nothing but that he will not be done out with cess, coyne nor liverie."[4] During the sixteenth century the proportion of this non-land-owning class in the population grew in comparison with the land-owning class; for instance, the population of the planted part of Wexford was 15,000, whereof no

[1] 10 & 11 Car. I., c. 16.   [2] Dr. Sigerson, *History of Irish Land Tenures.*
[3] See Joyce, *Smaller Social History of Ancient Ireland,* p. 80.
[4] Col. Thos. Smyth's *Settlement in the Ards,* 1572; *Ulster Journal of Arch.,* vol. ix., p. 177; and see Payne's *Description of Munster in* 1589, *Irish Arch. Soc.,* vol. i.

more than 667 claimed freeholds.¹ If we allow that the average family contained five members, we can calculate that there were four landless men to each land-owner in a clan. The existence of slavery, or rather serfdom, in Ireland, even in the seventeenth century, is illustrated by a grant enrolled in 1602, whereby the Baron of Upper Ossory was granted certain townlands " with all demesne lands, wages, tithes, uplands, obventions, waters, fisheries, mines and rents, serfs, advowsons, waifs, strays, native men and women (nativos et nativas).''² One of the first acts of Chichester's government was a proclamation in 1605 abolishing serfdom in Ireland.³

These churls or cottiers, as they afterwards came to be called, were at all times the labourers who tilled the land in Ireland, and they remained throughout the many changes of the seventeenth century a constant factor of Irish life—the lowest stratum of Irish society. " The tempest that devastated the castle swept over the cabin," says Dr. Sigerson; and again, " It must be borne in mind that in the wars the humbler class of cultivators generally escaped the change and destruction that fell upon their superiors in station. The honey was too welcome not to secure the toleration of the working bees; the English and Irish combatants looked down on them as slaves and churls, unfit for fighting, but apt to cultivate land and cattle, disinclined for war and revolutions if not pressed into them by intolerable oppression; they remained even through the Cromwellian transplantations the one comparatively fixed element in Irish social history, a settled sub-stratum.''⁴

It is difficult to ascertain whether the standard of life amongst these humbler cottiers was higher at this period than at a later date. It certainly could not have been lower, because in the eighteenth century they were on a bare margin of existence, and it is probable that in the earlier part of the seventeenth century they enjoyed

---

¹ Butler, *Policy of Surrender and Regrant*, p. 103.   ² Cal. Pat. Rolls, Eliz., p. 599.
³ Printed in Bonn, vol. i., p. 394; see Butler, *Policy of Surrender and Regrant*, p. 102; *Confiscations in Irish History*, p. 37.
⁴ Sigerson, *History of Irish Land Tenures*, p. 36.

more material comforts than at a later date. Even so, their condition was one of poverty and hard work remunerated by a wretched return. We obtain occasional glimpses of them in the writings of the time. They lived in "booths made of boughs, kept with long strips of green turf instead of green canvas, run up in a few moments"; the purpose of which was to enable them to move more easily with their cattle, if they had any, or with their chief's cattle which they were accustomed to herd.[1] An English traveller in 1619 gives us a detailed picture of the condition of the cottiers:—"Their houses are advanced three or four yards high pavilion-like encircling, erected in a singular frame of smoke-torn straw, green, long picked turf and rain-dropping wattles. Their several rooms, halls, parlours, kitchens, barns and stables are all enclosed in one, and that one perhaps in the midst of a mire, where, in foul weather, they scarcely can find a dry spot whereupon to repose their heads, their shirts being woven of the wool or linen of their own nation, and their penurious food resembles their miserable condition."[2] "The Irish homes are the poorest cabins I have ever seen, planted in the middle of the fields and grounds which they farm and rent."[3] "The common sort never kill any cattle for their own use, being contented to feed all the year upon milk, butter, and the like, and do eat but little bread."[4]

(5) *The State of Agriculture.*

During this period the condition of agriculture in Ireland was proverbially backward, and was referred to as notorious by James I. in a speech in London :—"In contending to have a committee before you agree on a speaker you did put the plough before the horse so that it went untowardly like your Irish plough."[5] "I remember," wrote

---
[1] C.S.P., Ire., 1611-14, p. xix.
[2] Lithgow's *Tour in Ireland*, 1619; *Jl. of Cork Arch. Soc.*, vol. viii., p. 104.
[3] Brereton's *Travels in Ireland*, 1635.
[4] *Advertisements for Ireland*, MS., T.C.D. A list of the wages paid to various grades of agricultural labourers in 1608 and 1640 is printed in Appendix II.
[5] C.S.P., Carew, 1603-24, p. 290.

a traveller in Ireland in 1619, " I saw in Ireland's north parts two remarkable sights—the one was their manner of tillage—it is as bad husbandry as ever I saw among the wildest savages alive."[1] Sixteen years later Sir William Brereton noticed that " the soil was overtilled and wronged . . . . slothfully and improvidently ordered, much impaired, and yielding much less than if well husbanded."[2] Of course, the reasons of this were plain. During the Elizabethan wars the Irish were driven from their lands to the woods and mountains, and lived " by the milk of the cow without husbandry or tillage."[3] The Irish land tenure, moreover, did not conduce to a high state of husbandry, as the owners of a parcel of land were constantly changing, and there was no inducement offered to create permanent improvements. Possibly, Davies was prejudiced against the Irish tenures, which he always sought to displace, but one cannot help thinking that there is some truth in his allegation that the customs of Coyne and Livery had a bad effect on agriculture; " it made the land worse and the people idle."[4] The practice of " creaghting " was particularly fatal to agricultural improvement.[5] Even on the planted lands tenants suffered from being granted none but short leases, which did not give them sufficient security to make it worth their while to improve; " the tenants have not care to make any strong or defensible buildings or houses, to plant or to enclose."[6]

Sir John Davies alleges that he noticed an improvement in the standard of agriculture in 1613 as a result of the redistribution of the lands, although of course his statement on this subject must be taken with some reserve. " And thus we see how the greatest part of the possessions (as well of the Irish as of the English) in Leinster, Connaught and Munster are

[1] Lithgow's *Tour in Ireland, Cork Arch. Jl.*, vol. viii., p. 104.
[2] *Ulster Jl. of Arch.*, o.s. vol. vi., p. 121. A good description of Irish Agricultural methods is contained in *Advertisements of Ireland*, T.C.D. MSS.
[3] Davies' *Discoverie*, p. 663.   [4] Davies' *Discoverie*, p. 669.
[5] *Ulster Jl. of Arch.*, o.s. vol. vi., p. 120.
[6] C.S.P., Carew, 1589-1600, p. 208; *Discourse for Reformation of the Province of Ulster*, by Justice Saxey.

settled and secured since his Majesty came to the Crown; whereby the hearts of the people are also settled, not only to live in peace, but raised and encouraged to build, to plant, to give better education to their children, and to improve the commodities of their lands; whereby the yearly value thereof is already increased double of that it was within these few years, and is like daily to rise higher, till it amount to the price of our land in England."[1] " The grievances, however, of particular persons," wrote Carte, " did not prevent the general good intended to the kingdom by these plantations; in consequence of which lands were cultivated and greatly improved, the product and commodities of the country increased."[2] The change from the Brehon to the Feudal tenure was analogous in its results to the enclosure of the common lands in England—great advantage to husbandry at the expense of great hardship to individuals.[3]

The feature of Irish agriculture which struck English observers most was the custom of ploughing by the tail, which, in the early seventeenth century, universally prevailed. The plough used was the short Irish plough to which James I. referred, and this was usually drawn by five or six horses abreast with a man to every horse; no harness of any kind was employed, but the horses were attached to the plough direct by the tail.[4] The English were ever anxious to abolish this custom on humanitarian grounds, but they did not succeed until a much later date than that of which we are writing. In 1606 an Act of Council forbade this custom under penalty of the forfeiture of one garron for the first offence, two for the second offence, and the whole team for the third. This Order was not put into execution until 1611, when a commuted fine of ten shillings for every plough so offending was collected in some counties. In 1612 ten shillings per plough was thus levied throughout Ulster,

[1] *Discoverie*, p. 711; Strafforde's *Letters*, vol. ii., p. 65.
[2] Carte, vol. i., p. 51; Leland, vol. ii., p. 473; Pynnar's *Survey, passim.*
[3] Gibbins, *Industry in England*, p. 275; Cunningham, *Growth of English Industry and Commerce*, vol. ii., p. 555.
[4] *A New Description of Ireland, with the Disposition of the Irish*, by Barnaby Rich, London, 1610, p. 26.

and some idea of the prevalence of the custom may be gathered from the fact that £870 was thus raised. The Irish were quite undeterred, and did not cease to plough by the tail, in spite of these penalties, as they rightly or wrongly considered this method the most suitable for the rough and hilly ground with which they were accustomed to deal.[1] In 1611, amongst the Acts proposed by the Lord Deputy was one entitled "An Act for abolishing the custom of drawing plough cattle only fastened by their tails; blowing their milch cattle to make them give milk and the pulling of sheep,"[2] but this Act did not become law. In 1615 the House of Commons noticed that the prohibition of ploughing by the tail was of no effect, and petitioned that a corporal punishment should be instituted for the pecuniary one.[3] There seems to have been a good deal of connivance on the part of the collectors of the fines throughout the country, as in 1620 it was stated that the fine of 10s. was reduced to one of 2s. 6d.[4] In 1634 an Act was passed making it a misdemeanour to practise the "barbarous custom of ploughing, harrowing, drawing, and working with horses, mares, garrons, and colts by the tail."[5] In 1640 the Lords' representatives petitioned the King that the Lord Deputy should be permitted to exercise a dispensing power under the statute;[6] and the unpopularity of the statute may be further adduced from the fact that its repeal was made one of the conditions of peace between Charles I. and the Irish Confederates in 1646.[7] Other Irish customs looked on as particularly backward were those of burning corn in the straw, and pulling the wool off live sheep, and these customs were forbidden by statute in 1634.[8]

The commonest manure used was the dung of cows and horses mixed with straw; sheep's dung was used in some parts, and experiments were made in the use of pigeon's dung. The custom of manuring with ashes was

---

[1] C.S.P., Ire., 1611-14, p. 449; Pynnar's Survey. [2] C.S.P., Carew, 1603-24, p. 163.
[3] I.C.J., vol. i., p. 32. [4] C.S.P., Ire., 1615-25, p. 283. [5] 10 & 11 C..r. I., ch. 15.
[6] Carte, vol. i., p. 211.
[7] Bagwell, vol. i., p. 125. On this subject see an Article in the Ulster Jl. of Arch., o.s. vol. vi., p. 212. [8] 10 & 11 Chas. I., ch. 17.

not much practised in Ireland, but manuring with lime, which in the eighteenth century became the standard fertiliser in Ireland, was looked on as a recent innovation in 1650.[1]

On the whole, in spite of the backward condition of agriculture in many respects, Irish land improved in the period under review. In 1630, Lord Cork was struck by the gradually increasing value of land: "There is a marvellous change from the state of affairs which old inhabitants can remember; buildings and farms are improving, each man striving to excel others in fair building and good furniture, and in husbanding and closing and improving their land."[2] The lands in the Plantation counties, where a reasonable fixity of tenure was possessed by the tenants, were noticed to have specially improved,[3] and a general increase in the value of lands in all parts of the country was observed.[4] In 1635 land in some parts of the country were sold for twenty years' purchase.[5]

Although the quantity of pasture was so great, the quality of the cattle was low. Fynes Moryson thought the Irish cattle in general very small, and attributed this to the fact that they ate only by day, and were all brought into the bawns of the castle at night.[6] "The number of the cattle is very great; yet would these be far more in number did not general murrain every fourth year consume them through improvidence and ill-husbandry."[7] Boate noticed that "The Irish kine, sheep, and horses are a very small size"[8] On the other hand, the manuscript, "Advertisements for Ireland," from which we have so often quoted, states that "Ireland abounds with more cattle, and such as be of English breeds and kept in choice grounds after the English husbandry be as large as the English be." The old breed of Irish hobby horses was dying out at this time, and the only horses in Ireland, apart from the small garrons of which one reads so much,

---

[1] A full account of the manuring of Irish land is to be found in Boate, *Ireland's Natural History*, 1652, chaps. 11 and 12.
[2] C.S.P., Ire., 1625-32, p. 590.   [3] Carte, vol. i., p. 51; Prendergast, p. 49.
[4] C.S.P., Ire, 1633-47, p. 270.   [5] Brereton, *Travels in Ireland*, 1635.   [6] *Itin.*, p. 222.
[7] *Advertisements for Ireland*, MS., T.C.D.   [8] *Ireland's Natural History*, c. 10.

were English horses for the importation of which licenses were frequently granted to English settlers.[1]

(6) *Pasture and Tillage.*

Of course, it goes without saying that the area under tillage during this period was considerably less than the area devoted to pasture.[2] This was a necessary result of the sparse population which was left in the country after the ravages of the Elizabethan Wars, and of the fact that the native Irish were primarily addicted to using their lands for the rearing of cattle. As far as the country was tilled at all, it was for oats for horses, and oatmeal, and for barley for *aqua vitae*;[3] as the diet of the lower classes consisted almost entirely of the product of cattle, corn only playing a small part therein. Wheat was scarcely grown by the native Irish at all, but was introduced by the English settlers: "Con O'Neill was so right Irish that he cursed all his posterity in case they either learnt English or sowed wheat."[4] The growth of tillage was discouraged by the insecurity of tenure which was universally felt as the result of the confiscations, and of the change of the land system; and by the "incidents of taxation which pressed on under-tenants while the great lords escaped."[5]

But in spite of this discouragement and of the preponderance of pasture, there was, during the earlier part of the seventeenth century, a considerable amount of tillage in Ireland compared with what there was at a later date. In 1600 the garrison of Derry burnt O'Doherty's granary, which was said to contain corn worth £3,000; "when Sir William Fitzwilliam last came over to govern our realm was plentiful in all kinds of provisions as corn was little worth."[6] In 1611, corn was exported to Spain.[7] "The Irish have been so addicted to tillage that a Bristol ban barrel of barley was sold but for 18d. in the market

[1] C.S.P., Ire., 1633-47, p. 46. [2] C.S.P., Ire., 1615-25, p. 86.
[3] C.S.P., Ire., 1611-14, p. 20. [4] Speed's *Chronicles*, London, 1611, p. 837.
[5] C.S.P., Ire., 1625-32, p. 462. [6] C.S.P., Ire., 1600-1, p. 94.
[7] C.S.P., Carew, 1603-24, p. 138.

of Coleraine."¹ In the following year, corn was transported in great quantities to Spain.² "There is a great store of corn in the country."³ In 1614 Sir Oliver St. John wrote: "Great good will come to this kingdom by transporting cattle and corn hence into England, for this kingdom will be able to spare great quantities of both."⁴ "In the ancient plantations, especially the English Pale, there is yearly as much corn as a man would imagine they might supply half England and reserve sufficient for their own use, too."⁵

There seems to have been a falling off in the amount of tillage in Ireland about 1620, although it is difficult to ascertain the cause. In 1619 there were complaints that the Plantation lands in Ulster were not being sufficiently tilled. "Many English do not yet plough nor use husbandry, being fearful to stock themselves with cattle or servants for those labours. Neither do the Irish use tillage, for they are also uncertain of their stay. So by these means the Irish using grazing only and the English very little, and were it not for the Scotch, who plough in many places, the rest of the country might starve. By reason of this the British, who are found to take their lands at great rates, live by the greater rents paid to them by the Irish tenants who graze."⁶ In 1623 the Commissioners of Ireland reported that the decrease in the amount of coin and bullion in Ireland proceeded from the "leaving off of tillage for corn, whereas theretofore the plenty of grain was such that by exporting the same great quantity of coin and bullion was brought in."⁷ There was an actual shortage of grain in Ireland about 1628, but, in spite of this, licences were granted for the export of corn to England to meet the shortage there.⁸ In 1628 Sir William St. Leger had to import corn for his cattle from England, as there was none in Ireland.⁹ The quantity of tillage must have continued to decrease, as in 1634 a

¹ C.S.P., Ire., 1611-14, p. 226.
² Barnaby Rich, *Remembrance of the State of Ireland in* 1612; Proc. R.I.A., vol. 26, c. p. 125.   ³ C.S.P., Ire., 1611-14, p. 379.   ⁴ C.S.P., Ire., 1611-14, p. 502.
⁵ *Advertisements for Ireland*, MS., T.C.D.   ⁶ Pynnar's *Survey*.
⁷ C.S.P., Ire., 1615-25, p, 425.   ⁸ C.S.P., Ire., 1625-32, pp. 146 and 593; Townshend, *Life and Letters of the Great Earl of Cork*, p. 184.   ⁹ C.S.P., 1625-32, p. 414.

committee was appointed by the Irish House of Commons to draw up a bill to prohibit the conversion of arable land into pasture, but no such bill ever became law.[1] The exports of corn in 1641 were as follows[2]:—

Oats (qrs.)	2,459
Oatmeal (qrs.)	492
Rapeseed (qrs.)	1,036

(7) *The Woods.*

The Irish woods suffered extensively during the Elizabethan wars, when they were systematically destroyed as affording harbour to the Irish rebels, and the consequence was that, at the beginning of the seventeenth century, many parts of Ireland were noticeably denuded of timber. In 1609 one of the inducements held out to the citizens of London to undertake the plantation of Ulster was the great store of timber for shipbuilding, but in the following year the planters complained they found the woods of Ulster were much wasted.[3] The woods of Munster were also largely destroyed.[4]

Apart from the deliberate destruction of the woods, which was such a strongly marked feature of military policy, there were various other causes operating in the same direction to deplete the country of timber. One of these was the large export. In 1603 the timber for most of the galleys built in Scotland came from Ireland,[5] and in 1610 great quantities of Irish timber were bought by the West India merchants for exportation.[6] "The wood thence transported yields a great revenue yearly, the very pipe staves afford a very great custom yearly were it truly exacted, for that all Spain and most of France is supplied from Ireland."[7] In 1641 the following were the exports of timber[8]:—

Timber	334	Tons
Boards and planks	209	,,
Barrel staves	941	,,
Hogsheads staves	663	,,
Pipe staves	144	,,

[1] *I.C.J.*, vol. i., p. 82. [2] C.S.P., Ire., 1669-70, p. 54. [3] C.S.P., Ire., 1608-10, pp. 208 & 349.
[4] *Ib.*, p. 88. [5] C.S.P., Ire., 1601-3, p. 667. [6] C.S.P., Ire., 1608-10, p. 530.
[7] *Advertisements for Ireland*, MS., T.C.D. [8] C.S.P., Ire., 1669-70, p. 54.

The woods were also greatly wasted by the growing ironworks;[1] indeed, ironworks were expressly encouraged with the object of destroying the woods.[2] In addition, a great deal of timber was used in Ireland by the inhabitants for building and fuel; up to the middle of the seventeenth century timber was the usual material employed for building Irish houses;[3] and, until the woods began to sink to very small proportions, turf was not at all employed for heating. It was also the common practice to fence in corn-fields with timber stakes, which were wastefully burnt down as soon as the harvest had been reaped. Of course, the destruction of the woods was accelerated by the insecure tenure of land, the temptation being very great to the tenant for the time being to make all the profit he could by cutting and selling the timber on the lands. " Within these few years the woods were all wasted with ill-husbandry, for the natives and possessors of these woods, not observing any seasons nor dividing them into coppices, do always cut down the main timber in all times of the year for their private benefit."[4]

From time to time various remedies were proposed to hinder the destruction of the woods. In 1609 the King's woods were ordered to be used only for the building of the King's ships,[5] and the export of timber was frequently prohibited by Proclamation.[6] Instructions were also given that planting should be encouraged;[7] and in the Parliament of 1611 an Act was proposed for the preservation of timber, which, however, did not become law.[8] In 1652 Boate wrote as follows about the woods:—" In antient times, and as long as the land was in the full possession of the Irish themselves, all Ireland was very full of woods on every side, as evidently appeareth by the writings of Giraldus Cambrensis, who came into Ireland upon the first Conquest, in the company of Henry the Second,

---

[1] Gernon, *Descriptions of Ireland*, 1620.   [2] C.S.P., Ire., 1608-10, p. 419.
[3] Sigerson, *Irish Land Tenures*, p. 251.   [4] *Advertisements for Ireland*, MS.,T.C.D.
[5] C.S.P., Ire., 1608-10, p. 174.   [6] C.S.P., Ire., 1611-14, p. 250; C.S.P., Ire., 1615-25, p. 48.
[7] C.S.P., Ire., 1611-14, p. 369.   [8] *Ib.*, p. 250.

King of England, in the year of our Saviour eleven hundred seventy and one. But the English having settled themselves in the land, did by degrees greatly diminish the Woods in all the places where they were masters, partly to deprive the Theeves and Rogues, who used to lurk in the Woods in great numbers, of their refuge and starting-holes, and partly to gain the greater scope of profitable lands. For the trees being cut down, the roots stubbed up, and the land used and tilled according to exigency, the Woods in most parts of Ireland may be reduced not only to very good pasture, but also to excellent Arable and Meddow.

"Through these two causes it is come to pass in the space of many yeares, yea of some ages, that a great part of the Woods, which the English found in Ireland at their first arrival there, are quite destroyed, so as nothing at all remaineth of them at this time.

"And even since the subduing of the last great Rebellion of the Irish before this, under the conduct of the Earl of Tirone (overthrown in the last yeares of Queen Elizabeth by her Viceroy, Sir Charles Blunt, Lord Mountjoy, and afterwards Earl of Devonshire), and during this last Peace of about forty yeares (the longest that Ireland ever enjoyed, both before and since the coming in of the English) the remaining woods have very much been diminished, and in sundry places quite destroyed, partly for the reason last mentioned, and partly for the wood and timber itself, not for the ordinary uses of building and firing (the which ever having been afoot, are not very considerable in regard of what now we speak of) but to make merchandise of, and for the making of Charcoal for the Iron works. As for the first, I have not heard that great timber hath ever been used to be sent out of Ireland in any great quantity, nor in any ordinary way of Traffick; but only Pipe-staves, and the like, of which good store hath been used to be made, and sent out of the land, even in former times, but never in that vast quantity, nor so constantly as of late years, and during the last Peace, wherein it was grown one of the ordinary merchandable commodities of the countrie, so as a mighty Trade was driven in them, and whole shiploads sent into forrein countries yearly; which as it brought great profit to the proprietaries, so the felling of so many thousands of trees every year as were employed that way, did make a great destruction of the Woods in tract of time. As for the Charcoal, it is incredible what quantity thereof is consumed by one Iron-work in a year; and whereas there was never an Iron-work in Ireland before, there hath been a very great number of them erected since the last Peace in sundrie parts of every Province: the which to furnish constantly with Charcoales, it was necessary from time to time to fell an infinite number of trees, all the lopings and windfals being not sufficient for it in the least manner.

" Through the aforesayd causes Ireland hath been made so bare of Woods in many parts, that the inhabitants do not only want wood for firing (being therefore constrained to make shift with turf, or sea-coal, where they are not too far from the sea), but even timber for building, so as they are necessitated to fetch it a good way off, to their great charges, especially in places where it must be brought by land: And in some parts you may travell whole dayes long without seeing any woods or trees except a few about Gentlemen's houses; as namely from Dublin, and from places that are some miles further to the South of it, to Tredagh, Dundalke, the Nurie, and as far as Dremore; in which whole extent of land, being above three-score miles, one doth not come neer any woods worth the speaking of, and in some parts thereof you shall not see so much as one tree in many miles. For the great Woods which the Maps do represent unto us upon the Mountains between Dundalke and the Nurie, are quite vanished, there being nothing left of them these many yeares since, but one only tree, standing close by the highway, at the very top of one of the Mountains, so as it may be seen a great way off, and therefore serveth travellers for a mark.

IN THE SEVENTEENTH CENTURY. 47

"Yet notwithstanding the great destruction of the Woods in Ireland, occasioned by the aforesayd causes, there are still sundry great Woods remaining, and that not only in the other Provinces, but even in Leinster itself. For the County of Wickloe, Kings's-county, and Queens-county, all three in that Province, are throughout full of Woods, some whereof are many miles long and broad. And part of the Counties of Wexford and Carlo are likewise greatly furnished with them.

"In Ulster there be great Forests in the County of Doneghall, and in the North part of Tirone, in the Country called Glankankin. Also in the County of Fermanagh, along Lough-Earn; in the County of Antrim; and in the North-part of the County of Down; in the two Countries called Killulta and Kilwarlin; besides several other lesser Woods in sundrie parts of that Province. But the County of Louth, and far the greatest part of the Counties of Down, Armagh, Monaghan, and Cavan (all in the same Province of Ulster) are almost every where bare, not only of Woods, but of all sorts of Trees, even in places which in the beginning of this present Age, in the War with Tirone, were encumbred with great and thick Forests.

"In Munster where the English, especially the Earl of Cork, have made great havock of the Woods during the last Peace, there be still sundrie great Forests remaining in the Counties of Kerry, and of Tipperary; and even in the County of Cork, where the greatest destruction thereof hath bin made, some great Woods are yet remaining, there being also store of scattered Woods both in that County, and all the Province over.

"Connaught is well stored with trees in most parts, but hath very few Forests or great Woods, except in the Counties of Mayo and Sleigo."[1]

(8) *The Mines.*

The large mineral wealth of Ireland was not made much use of during this period. In 1605 a licence

---
[1] Boate, *Natural History of Ireland*, chap. xv.

was granted to dig a mine of sea coal "near the river of Shannon in Munster," but it is not clear what mine is referred to.[1] This is the only reference to coal-mining until Boate refers to the mines in County Carlow. "As for Sea-Coales, they are the ordinary firing in Dublin, and in other places lying near the sea, where the same in time of peace are brought in out of England, Wales, and Scotland, in great abundance, and therefore reasonable cheap: which is the reason, that the less care hath been taken to find out Coal-mines in Ireland itself, whereas otherwise it is the opinion of persons knowing in these matters, that if diligent search were made for them, in sundry parts of the land, good Coal-mines would be discovered. This opinion is the more probable, because that already one Coal-mine hath bin found out in Ireland, a few yeares since, by meer hazard, and without having been sought for. The Mine is in the Province of Leinster, in the County of Carlo, seven miles from Idof, in the same hill where the Iron-mine was of Mr. Christopher Wandsworth, of whom hath been spoken above. In that Iron-mine, after that for a great while they had drawn Iron-oar out of it, and that by degrees they were gone deeper, at last in lieu of Oar they met with Sea-coal, so as ever since all the people dwelling in those parts have used it for their firing, finding it very cheap; for the load of an Irish-car, drawn by one garron, did stand them, besides the charges of bringing it, in nine-pence only, threepence to the digger, and sixpence to the owner.

"There be coales enough in this Mine for to furnish a whole country; nevertheless, there is no use made of them further than among the neighbouring inhabitants; because the Mine being situated far from Rivers, the transportation is too chargeable by land.

"These Coals are very heavy, and burn with little flame, but lye like Char-coal, and continue so the space

[1] Rep. Pat. Rolls, Jas. I., vol. i., pt. i., p. 192.

IN THE SEVENTEENTH CENTURY. 49

of seven or eight houres, casting a very great and violent heat."[1]

There was a good deal of iron-work done during this period. Extensive mining and smelting works were in operation from early in the century at Arigna and other places in County Leitrim;[2] at Toome in Ulster, in 1609, we read of an Irish smith of extraordinary skill who, in the opinion of an English traveller, was able to perform feats quite unequalled by any smith whom he had met elsewhere;[3] and the native Irish were skilful workers of iron, which they made into skenes and darts.[4] In Munster there were many ironworks which helped to exhaust the woods of the country; in 1626 we hear of cannon being cast at Cappoquin;[5] and the making of ordnance in Ireland on an extensive scale was contemplated.[6] The great Earl of Cork was ever zealous in his encouragement of the iron-works. "He sunk mine after mine; at Ballyregan, Cappoquin, Mocollop, Ardglyn, Kilmacoe and Lisfinnon, the forges glowed till the Blackwater valley bid fair to be an Irish black country."[7] He also attempted iron-works at Scariff and in County Leitrim.[8] Iron-works were encouraged by the Government; in 1610, £3,000 was granted for this purpose,[9] but, unfortunately, one of the reasons why iron-works were encouraged was to denude the country of timber.[10] There were exported from Ireland in 1641, 778 tons of wrought iron;[11] and the neighbourhood of Belfast was specially rich in forges.[12]

Towards the middle of the century, Boate gives a long account of the iron-works: "But to let alone uncertain conjectures, and to content ourselves with the Mines that are already discovered, we will in order speak of them, and begin with the Iron-mines. Of them there are three sorts in Ireland, for in some places the Oar of the Iron is drawn out of Moores and Bogs, in others it is hewen

---
[1] Boate, *Ireland's Natural History*, ch. xix., pp. 123-4.
[2] *Journal of the R. S. of Antiquaries*, Ireland, vol. xxxvi., p. 128.
[3] C.S.P., Ire., 1608-10, p. 290.   [4] C.S.P., Ire.. 1611-14, p. 227.
[5] C.S.P., Ire., 1625-32, pp. 110 and 314.   [6] C.S.P., Ire., 1647-60, p. 74.
[7] Townshend, *Life and Letters of the Great Earl of Cork*, p. 102.
[8] *Ibid.*   [9] C.S.P., Ire., 1608-10, p. 432.   [10] *Supra* page 44.
[11] C.S.P., Ire., 1669-70, p. 55.   [12] Benn, *History of Belfast*, p. 334.

out of Rocks, and in others it is digged out of Mountains: of which three sorts the first is called Bog-Mine, the other Rock-Mine, and the third with severall names White-mine, Pin-mine, and Shel-mine.

"The first sort, as we have said, and as the name it self doth shew, is found in low and boggie places, out of the which it is raised with very little charge, as lying not deep at all, commonly on the superficies of the earth, and about a foot in thickness. This Oar is very rich in metall, and that very good and tough, nevertheless in the melting it must be mingled with some of the Mine or Oar of some of the other sorts: for else it is too harsh, and keeping the furnace too hot it melteth too suddenly, and stoppeth the mouth of the furnace, or, to use the workmen's own expression, choaketh the furnace. Whilest this Oar is new, it is of a yellowish colour, and the substance of it somewhat like unto clay, but if you let it lye any long time in the open air, it groweth not only very dry, as the clay useth to do, but moldereth and dissolveth of it self, and falleth quite to dust or sand, and that of a blackish or black-brown colour.

"The second sort, that which is taken out of Rocks, being a hard and meer stony substance, of a dark and rustie colour, doth not lye scattered in several places, but is a piece of the very rock, of the which it is hewen; which rock being covered over with earth, is within equallie every where of the same substance; so as the whole Rock, and every parcell thereof, is Oar of Iron. This mine, as well as the former, is raised with little trouble, for the Iron-rock being full of joints, is with pick-axes easily divided and broken into pieces of what bigness one will: which by reason of the same joints, whereof they are full every where, may easily be broke into other lesser pieces; as that is necessary, before they be put into the furnace.

"This Mine or Oar is not altogether so rich as the Bog-mine, and yeeldeth very brittle iron, hardly fit for any thing else, but to make plow-shares of it (from whence the name of coltshare Iron is given unto it), and therefore is seldom melted alone, but mixed with the first or the third sort.

"Of this kind hitherto there hath but two Mines been discovered in Ireland, the one in Munster, neer the town of Tallo, by the Earl of Cork his Iron works; the other in Leinster, in Kings-county, in a place called Desert land, belonging to one Serjeant Major Piggot, which rock is of so great a compass, that before this rebellion it furnished divers great Iron-works, and could have furnished many more, without any notable diminution; seeing the deepest pits that had been yet made in it, were not above two yards deep. The land, under which this rock lyeth, is very good and fruitfull, as much as any other land thereabouts, the mold being generally two feet and two and a half, and in many places three feet deep.

"The third sort of Iron-mine is digged out of the mountains, in several parts of the Kingdome; in Ulster, in the County of Fermanagh, upon Lough Earn; in the County of Cavan, in a place called Douballie, in a drie mountain; and in the County of Nether-Tirone, by the side of the rivelet Lishan, not far from Lough Neaugh; at the foot of the mountains Slew-galen mentioned by us upon another occasion, in the beginning of this chapter; in Leinster, in Kings-county, hard by Mountmellick; and in Queenes-county, two miles from Monrath; in Connacht; in Tomound of the County of Clare, six miles from Limmerick; in the County of Roscomen, by the side of Lough Allen; and in the County of Leitrim, on the Eastside of the said Lough, where the mountains are so full of this metall, that thereof it hath got in Irish the name of Slew Neron, that is, Mountain of Iron; and in the province of Munster also in sundrie places.

"This sort is of a whitish or gray colour, like that of ashes; and one needs not take much pains for to find it out, for the mountains which do contain it within themselves, do commonly show it of their own accord, so as one may see the veins thereof at the very outside in the sides of the mountains, being not very broad, but of great length, and commonly divers in one place, five or six ridges the one above the other, with ridges of earth between them.

"These Veins or Ridges are vulgarly called Pins, from whence the Mine hath the name of Pin-mine; being also called White-mine, because of its whitish colour; and Shell-mine, for the following reason: for this stuff or Oar being neither loose or soft as earth or clay, neither firm and hard as stone, is of a middle substance between both, somewhat like unto Slate, composed of shells or scales, the which do lye one upon another, and may be separated and taken asunder very easily, without any great force or trouble. This stuff is digged out of the ground in lumps of the bigness of a man's head, bigger, or less, according as the Vein affordeth opportunitie. Within every one of these lumps, when the Mine is very rich and of the best sort (for all the Oar of this kind is not of equall goodness, some yeelding more and better Iron than other) lieth a small kernell which hath the name of Hony-comb given to it, because it is full of little holes, in the same manner as that substance whereof it borroweth its appellation.

"The Iron comming of this Oar is not brittle, as that of the Rock-mine, but tough, and in many places as good as any Spanish Iron.

"The English having discovered these Mines, endeavoured to improve same, and to make profit of them, and consequently severall Iron-works were erected by them in sundrie parts of the land, as namely by the Earl of Cork in divers places in Munster; by Sir Charles Coot in the Counties of Roscommon and Letrim, in Connacht, and in Leinster by Monrath, in Queenes-county; by the Earl of Londonderry at Ballonakill, in the sayd county; by the Lord Chancelour Sir Adam Loftus, Viscount of Ely, at Mountmelik, in Kings-county; by Sir John Dunbar in Fermanagh, in Ulster; and another in the same county, by the side of Lough-Earn, by Sir Leonard Blenerhassett; in the county of Tomong, in Connacht, by some London-

Merchants; besides some other Works in other places, whose first erectors have not come to my knowledge.

"In imitation of these have also been erected divers Iron-works in sundrie parts of the sea coast of Ulster and Munster, by persons, who having no Mines upon or near their own Lands, had the Oar brought unto them by sea out of England; the which they found better cheap than if they had caused it to be fetched by land from some of the Mines within the land. And all this by English, whose industrie herein the Irish have been so far from imitating, as since the beginning of this Rebellion they have broke down and quite demolished almost all the forementioned Iron-works, as well those of the one as of the other sort."[1]

There was also an important silver and lead mine in County Tipperary, which was first worked in 1617,[2] although it had been discovered many years previously.[3] We read of another silver mine being discovered about 1630, although it is not clear in what part of the country;[4] and the Earl of Cork worked silver and lead mines at a place called Minehead, east of Youghal.[5] There was a grant in 1625 by the King of gold and silver mines in County Kerry,[6] which were extensively worked.[7] In 1641 over two hundred tons of lead were exported from Ireland, probably from the Tipperary mines.[8] The mines in Tipperary were granted by patent to Sir George Hamilton, who laid out large sums on working them, so that the profits which the King derived from his patent in 1640 were over £800 a year; over five hundred Englishmen were given employment in the works, and also many of the Irish and foreigners ; but these works were completely destroyed during the Rebellion of 1641.[9] Boate gives the following account of the silver mines :—

"Mines of Lead and Silver in Ireland have to this day been found out, three in number; one in Ulster, in

---

[1] Boate, *Natural History of Ireland*, ch. 16.   [2] C.S.P., Ire., 1615-25, p. 195.
[3] C.S.P., Ire., 1611-14, p. 251.   [4] *Franciscan MSS.*, p. 35.
[5] Townshend, *Life and Letters of the Great Earl of Cork*, p. 105.
[6] C.S.P., Ire., 1625-32, p. 686.   [7] Cooper, *Life of Wentworth*, vol. i., p. 181.
[8] C.S.P., Ire., 1669-70, p. 55.   [9] C.S.P., Ire., 1660-2, p. 153.

the County of Antrim, very rich, forasmuch as with every thirty pounds of Lead it yieldeth a pound of pure Silver; another in Connacht, upon the very Harbour-mouth of Sligo, in a little Demy-Illand commonly called Conny-Illand; and a third in Munster. The first two having been discovered but a few years before this present Rebellion, were through severall impediments never taken in hand yet; wherefore we shall speak only of the third.

"The Mine standeth in the County of Tipperary, in the Barony of Upper Ormond, in the Parish of Kilmore, upon the Lands of one John MacDermot O-kennedy, not far from the Castle of Downallie, twelve miles from Limerick, and three-score from Dublin. The Land where the Mine is, is mountainous and barren; but the bottoms, and the lands adjoyning, are very good for Pasture and partly Arable; of each whereof the Miners had part, to the value of twenty pounds sterling per annum, every one. It was found out not above forty years ago, but understood at the first only as a Lead-mine, and accordingly given notice of to Donogh Earl of Thomond, then Lord President of Munster, who made use of some of the Lead for to cover the house which he then was building at Bunrattie; but afterwards it hath been found, that with the Lead of this Mine there was mixed some Silver.

"The Veins of this mine did commonly rise within three or four spits of the superfices, and they digged deeper as those Veins went, digging open pits very far into the ground, many fathoms deep, yea Castle-deep; the pits not being steep, but of that fashion as people might go in and out with wheel-barrows, being the only way used by them for to carry out the Mine or Oar. The water did seldom much offend them; for when either by the falling of much rain, or by the discovering of some Spring or Water-source, they found themselves annoyed by it, they did by Conduits carry it away to a brook adjoyning, the Mountain being so situate, as that might be done easily.

"This Mine yeelds two different sorts of Oar; of which the one, and that the most in quantitie, is of a reddish colour, hard and glistening; the other is like a Marle, something blewish, and more soft than the red; and this was counted the best, producing most Silver, whereas the other, or glistening sort, was very barren, and went most away into litteridge or dross.

"The Oar yeelded one with another three pound weight of Silver out of each Tun, but a great quantitie of Lead so as that was counted the best profit to the farmer.

"Besides the Lead and Silver the Mine produced also some Quicksilver, but not any Alome, Vitriol, or Antimony, that I could hear of.

"The Silver of this Mine was very fine, so as the Farmers sold it at Dublin for five shillings two pence sterling the ounce; as for the Lead, that they sold on the place for eleven pounds sterling the Tun, and for twelve pounds at the City of Limerick. The King had the sixth part of the silver for his share, and the tenth part of the Lead, the rest remaining to the farmers, whose clear profit was estimated to be worth two thousand pounds sterling yearly.

"All the Mills, Melting-houses, Refining-houses, and other necessary Work-houses, stood within one quarter of a mile at the furthest from the place where the Mine was digged, every one of them having been very conveniently and sufficiently built and accommodated by the Officers and substitutes of Sir William Russell, Sir Basil Brook, and Sir George Hamilton, which three persons successively had this Mine in farm from the King, but in the beginning of this present Rebellion all that hath been destroyed by the Irish under the conduct of Hugh O-kennedy, brother of John Mac-Dermott O-kennedy, on whose lands the Mine was situated; which Rebels not content to lay waste the mine, and to demolish all the works thereunto belonging, did accompany this their barbarousness with bloody cruelty against the poor workmen, such as were imployed about the melting and refining of the Oar, and in all offices thereunto belonging; the which some of them being English, and the rest Dutch (because the Irish having no skill at all in any of those things, had never been imployed in this mine otherwise than to digg it and to do other labours) were all put to the sword by them, except a very few who by flight escaped their hands."[1]

Copper mining was unimportant, in spite of the efforts of the Earl of Cork to encourage it.[2] In 1639 the Governor and Company of the Copper mines in England were granted a monopoly of working the copper ore in Ireland,[3] but the outbreak of the Rebellion prevented any use being made of the grant.

There were no gold mines worked in Ireland at this time, although there was a general belief in their existence. The grant we have referred to of the silver mines in Kerry also contains a reference to gold mines in that county; and again in 1623 we read that many mines of all sorts of metal " are being daily discovered and set on work."[4] Boate refers to the existence of gold deposits in County Tyrone :—

" I believe many will think it very unlikely, that there should be any Gold-mines in Ireland; but a credible person hath given me to understand, that one of his acquaintance had severall times assured him, that out of a certain rivelet in the County of Nether Tirone, called Miola (the which rising in the Mountains Slew-galen, and passing by the village Maharry, falleth into the Northwest corner of Lough Neaugh, close by the place where the river Band commeth out of it) he had gathered about one dram of pure gold; concluding thereby, that in the aforesayd Mountains rich Gold mines do lye hidden.

" For it is an ordinary thing for Rivers, which take their originall in gold-bearing mountains, to carry Gold mixt with their sand; the which may be confirmed by many instances, and to say nothing of severall rivers of

---
[1] Boate, *Natural History of Ireland*, ch. 18.
[2] Townshend, *Life and Letters of the Great Earl of Cork*, p. 105.
[3] C.S.P., Dom., 1690-1, pp. 91, 486, 498. [4] C.S.P., Ire., 1615-25, p. 426.

that kind, mentioned by Strabo, Pliny, and other old Geographers and Historians, nor of Pactolus and Hermus in Lydia, and Tagus in Spain, whereof all the old Poets are full; it is certain, that in our very times severall rivers in Germanie, as the Elbe, Schwarts, Sala, and others, do carry gold, and have it mixed with their sands; out of the which by the industry of man, it is collected."[1]

(9) *Trade*.

The distinguishing feature of Irish trade during the first forty years of the seventeenth century was its freedom from restraint. During the reign of Elizabeth a systematic war had been waged against Irish trade, and, as we have seen above, the last few years of the sixteenth century constituted a period of complete stagnation. After the Restoration, moreover, Irish trade was hampered by severe restrictions; the trade with England was almost entirely paralysed by the Cattle Acts and the high import duties imposed by the English Parliament; intercourse with the Colonies was impeded by the Navigation Acts; and ultimately even foreign trade was stifled by English legislation. Although both the reigns of Elizabeth and Charles II. were periods when attacks were made on Irish trade, the nature of the attacks was fundamentally different. In the former reign the aim was to divert Irish trade from the hands of the Irish into those of the English inhabitants of Ireland, who had the Queen's interests at heart, and who would use their prosperity and maritime skill to further her interests in Ireland; in the latter reign, on the other hand, the aim was to prevent Irish trade from interfering with or rivalling that of England. The one attack was made on certain individuals in Ireland in favour of others; the other on Irish traders as a whole. The period between the accession of James I. and the Restoration was comparatively favourable to the development of Irish trade. Although there were restraints

[1] Boate, *Natural History of Ireland*, chap. 18.

on the export of certain important commodities, these restraints were imposed by the Irish Parliament, and could therefore be repealed, subject to Poyning's Act, by the same body that created them. These restraints, moreover, were conceived with the object of aiding Irish industry, and in this respect were markedly different from the restraints of later years, which were designed with the express purpose of crippling the industrial life of the country.

The comparatively favoured position of Ireland during these years may be explained in the first place by the fact that the Stuarts were engaged in their great conflict with the English Parliament, and were greatly concerned to raise a revenue from Ireland, which would put them in a position to govern without the necessity of resorting to Parliament for supplies; and in the second place by the fact that Ireland was still too poor and undeveloped to appear as a serious competitor to England.

At the beginning of the seventeenth century, then, there was practically no restraint on the freedom of Irish trade, with the exception of certain prohibitions on exportation imposed by the Irish Parliament. Amongst these was a prohibition on the export of corn when the price was more than tenpence the peck,[1] and the Acts of Elizabeth placing high duties and penalties on the export of wool, flock, flax, linen, yarn, woollen yarn, sheepfell, calfsfell, goatfell, deerfell, beef-tallow, wax and butter, except under certain stringent conditions.[2] The Act of Henry VIII. completely forbidding the export of raw wool was obsolete in practice.[3] These restraints, however, were more a matter of theory than of practice, because licences to export great quantities of the prohibited commodities were frequently granted. They were also evaded by means of smuggling, and by the extensive trade with pirates that was carried on.[4] Indeed, the greater part of the trade of Ireland was carried on in these prohibited com-

[1] 12 Edward IV. [2] 11 Eliz., Sess. 3, c. 10.
[3] Irish Statutes under Tit. Wool. [4] C.S.P., Ire., 1615-25, p. 580.

modities; large quantities of corn, tallow, fells and wax were sent to Spain, and of yarn to England.[1]

Smuggling was not at that time considered so serious an offence against the Royal revenue as it became in the later period of high protective duties, the import and export duties in Ireland being very small.[2]

The statesmen of the reign of James I. did not altogether approve of the extent to which Ireland traded with Spain, but the objection was political rather than mercantile. It was felt that the constant intercourse between Irish and Spanish Catholics which the trade occasioned tended to keep alive the grievances of the Irish Catholics,[3] and that, if the Irish could be induced to trade more with England, the King would have the advantage of double customs.[4] However, no serious attempt was made to hinder the Irish-Spanish trade; possibly because the later theory that the English Parliament should regulate Irish trade had not then developed.

A certain check had been imposed on the normal progress of trade by an Act of 1495,[5] which provided that no Irishman should trade with another, but only with an Englishman; but this Act was repealed in 1614.[6] How far the repeal of the Act was carried into actual practise is a matter of some doubt, for as late as 1641 the fact that one Irishman was forbidden to trade with another was complained of as a grievance then felt.[7]

Trade was also discouraged by the "overruling privileges of most of the port towns," which claimed the right to exact exorbitant dues, and to exercise a right of pre-emption of imported commodities in favour of the townsmen.[8] James I. endeavoured to remedy this evil by granting new charters to the maritime towns, restricting their rights in these respects,[9] but the evil practices of the towns seem to have survived, for as late as 1632 complaints were made that "great decay of trade is caused

[1] C.S.P., Carew, 1603-24, pp. 84, 174.
[2] They were fixed by 15 Henry VII., c. 1, at twelvepence in the pound on all commodities imported or exported. [3] C.S.P., Carew, 1589-1600, pp. 458-9, ; 1601-3, p. 388.
[4] C.S.P., Carew, 1601-3, p. 436. [5] 10 Henry VII. [6] 11, 12 & 13 Jac. I., c. 5.
[7] C.S.P., Ire., 1633-47, p. 321. [8] C.S.P., Ire., 1606-8, p. 74.
[9] Davies' *Discoverie*, p. 712.

by these exactions ";[1] and in 1635 complaints were made in Parliament of the excessive tolls in fairs and markets.[2] James I. also endeavoured to stimulate trade in the country parts of Ireland by the erection of non-maritime corporate towns, and the establishment of cities and fairs.[3] These charters usually made it a condition that nobody should trade in a town unless he was a freeman, but this rule was universally neglected, with very bad results to the growth of industry. "The ancient corporate system was meant to benefit trade—as a matter of fact, it worked simply to the destruction of trade and commerce, and in favour of the political aggrandisement of some individuals."[4]

Trade also suffered from the existence of monopolies in many articles, amongst others starch, glass, china, and soap. " The country much complains of the monopolies there conferred on private men to the King's slender benefit, and which be of as great value if they were duly exacted, as the Crown revenue there."[5] In 1628 the monopolies of selling wine and aqua vitae which had been recently granted were admitted by the King to be oppressive,[6] but the system was extended by Wentworth, who granted a monopoly of tobacco, one of the most important articles of consumption in Ireland,[7] and who made attempt to create a monopoly of salt, which would put the Government in a position completely to control the Irish provision and fishing industries.[8] In 1640 the House of Commons presented a remonstrance against the "universal and unlawful increasing of monopolies, to the advantage of the few, and to the impoverishment of the people," and stated that there existed monopolies of tobacco, alum, iron pots, starch, glass, tobacco pipes, gunpowder, wine, and aqua vitae,[9] and in

---

[1] Egmont MSS., pp. 70-1. [2] *I.C.J.*, vol. i., p. 114; C.S.P., Dom., 1672-3, p. 84.
[3] Davies' *Discoverie*, p. 712.
[4] Stokes, *Athlone in the Seventeenth Century*; J.R.S.A.I., vol. xxi., pp. 198-9.
[5] *Advertisements for Ireland*, MS., T.C.D.   [6] C.S.P., Ire., 1625-32, p. 331.
[7] C.S.P., Ire., 1633-47, p. xiv.
[8] Strafforde's proposal in this matter is interesting, but it is too long to insert in the text; it is printed in Appendix III.
[9] *I.C.J.*, vol. i., p. 163; C.S.P., Ire., 1633-47, p. 313.

IN THE SEVENTEENTH CENTURY.

reply the King declared that all monopolies except those of alnage and glass should be abolished.[1]

Trade was also discouraged by the great variety of measures which were used in different parts of the country;[2] by the corruption and swindling of the customhouse officers;[3] and by the practice of granting and revoking arbitrary prohibitions and licences. For instance, on the 27th of June, 1614, a proclamation was made allowing the free export of prohibited commodities; this was revoked on the 7th of September in the same year; and again put into force on the 18th of December.[4] This arbitrary and unsatisfactory procedure was emphatically protested against by Chichester, who pronounced it "ruinous to trade."[5] Foreign merchants were alleged to be discouraged from setting up business in Ireland. "No foreign factors there reside by reason of the rare intercourse of traffic between them and the home merchants; the native merchants there bar all the foreign merchants, and likewise the very British from trading there within their town liberties, unless they have their commodities at their own price; and if they take any lodging or house from the native merchants they ruin their house rents extremely, and lay extraordinary town impositions and heavy taxes upon them to weary them out; which discourages all strangers to traffic thither."[6] But the thing that, above all others, had a depressing effect on the Irish foreign trade was the debasement of the coinage. This is dealt with more fully in a later section.

In 1625 war broke out between England and Spain, and this had an injurious effect on Irish trade, as Spain was the principal market for many Irish commodities. In 1626 a further restraint was put on trade by a proclamation prohibiting the export of Irish cattle to any part of the world, as the stock was becoming depleted by excessive exportation.[7] In the following year an embargo was

---

[1] Carte, vol. i, p. 283; on monopolies during the same period in England see Hewins, *English Trade and Finance.*
[2] C.S.P., Ire., 1608-10, p. 261. [3] C.S.P., Ire., 1606-8, p. 74.
[4] C.S.P., Ire., 1611-4, p. 502. [5] C.S.P., Ire., 1611-4, p. 535.
[6] *Advertisements for Ireland*, MS., T.C.D. [7] C.S.P., Ire., 1625-32, p. 159.

placed on the export of cows, beef, corn, tallow, leather, hides and butter, except to England.¹ This embargo caused much resentment, but a further blow was struck in the same year by the revision of the Book of Rates, which meant that every article was bound to pay a higher Customs duty than before. The following were the principal changes made in the rates.²

Commodity.	Old Rate.			New Rate.		
	£	s.	d.	£	s.	d.
Cattle	0	1	0	0	2	0
Wool (per stone)	0	0	2½	0	0	4
Yarn (per pack)	0	13	4	1	0	0
Tallow (per cwt.)	0	1	0	0	1	6
Butter (per barrel) ...	0	1	4	0	2	0
Herrings (per barrel) ...	0	0	9	0	1	0
Pilchards (per ton) ...	0	4	0	0	6	0
Rugs (per score)	0	10	0	0	12	0
Salt (per weigh)	0	1	4	0	2	0
Wines of France (per butt) ...	—			1	6	8
Wines of France (per tun) ...	—			1	13	4
Hides, into foreign parts ...	0	0	6	0	1	0

The many restraints on exportation must have had a very injurious effect on the trade of the Kingdom, and their removal was one of the clauses of the Graces conceded by the King in 1628. " For the advancement of trade and bringing of coin into the Kingdom, linen yarn, wool, corn, pipestaves, and living cattle were allowed to be transported into England without licence; and tallow, hides, fish, beef and pork in cask might be transported into any of the King's dominions and all other States in unity with him."³ It is well known how Charles I. succeeded in evading the performance of his promised " Graces," and it is unnecessary to add that the relaxations of the trade restraints were not confirmed any more than were any of the other articles of the compact.⁴ The restraints, therefore, continued, and much depression and loss were suffered in consequence.⁵

¹ C.S.P., Ire., 1625-32, p. 281.   ² Wood MSS., p. 192.   ³ Carte, vol. i.,p. 105.
⁴ Walpole, *The Kingdom of Ireland*, pp. 213-14.   ⁵ C.S.P., Ire., 1625-32, p. 469.

The hordes of pirates, moreover, that infested the southern coast also had a depressing effect on trade.[1] Wentworth pointed out, when dealing with the King's intentions with regard to the Graces, that the provisions relating to free export were not fit subjects for confirmation by Act of Parliament, and that any concessions made in that respect were only fit to be continued as instructions during the King's good pleasure.[2] Charles I. was, however, not actually hostile to Irish trade; on the contrary, he directed the Lords Justices to encourage it by every means in their power,[3] and Commissioners of Trade, one for each province, were appointed to attend the Privy Council to advise it on matters of trade and commerce.[4] In 1634 a grand committee of the House of Commons met to consider and advise on the erection and setting up of trade and manufactures.[5] It must also be remembered that at this time England also suffered from many restraints on the liberty of its trade, many of them arbitrary; indeed, the trade of Ireland was, if anything, freer than that of England.[6]

Wentworth was anxious to encourage Irish trade, as he wished to increase the King's Irish revenue, so as to render him less dependent for financial support on the English Parliament. One of the first projects he conceived on taking over the government of Ireland was an arrangement with the King of Spain that the Spanish fleet should be victualled with Irish provisions; but although negotiations with Spain were opened up, the project fell through.[7] A petition of the House of Commons that the ports should be freely opened for the export of all commodities, save wool and woolfells, was refused.[8] Wentworth, however, "complained in the Council of England that the Irish were treated in many cases as foreigners and instanced particularly in the imposition upon coals, whereon they paid four shillings a ton, as much as either the French or Dutch paid; in the excessive rate set upon horses transported

[1] C.S.P., Ire., 1625-32, p. 645 ; Strafforde's *Letters, passim*. [2] Carte, vol. i, p. 162.
[3] C.S.P., Ire., 1625-32, p. 472. [4] C.S.P., Ire., 1615-25, p. 402. [5] *I.C.J.*, vol. i., p. 67.
[6] Egmont MSS., p. 71. [7] Strafforde's *Letters*, vol. i., p. 93. [8] *I.C.J.*, vol. i., p. 69.

out of England into Ireland, where they wanted a good breed, and had not enough to supply even the occasions of the Army ; and in the eighteenpence set upon every live beast that came from thence. This representation caused all those duties to be taken off."[1] Again in 1637, when a proposal was made to the King that the Book of Rates should be raised, Wentworth remonstrated strongly against any such step, for fear that enhancing the rates should discourage the merchants and prejudice the commerce of the Kingdom.[2] Wentworth was extremely anxious to remove the restraints imposed on Irish tallow, but the opposition in England was obdurate, and the restraint remained in spite of the Lord Deputy's earnest appeals.[3] The most signal benefit conferred by Wentworth on the trade of Ireland was his clearing the sea of pirates. As early as 1635 it was said that "pirates are banished and trade reviving," and in the end the sea was completely cleared.[4]

The result of Wentworth's efforts to improve trade may be judged from the following figures, which show the Customs Duties collected for each year from 1632-1640,[5] but it must be remembered that the large increase in customs here shown was to some extent caused by the increase in the export to England of raw wool, the increased importation of English cloths,[6] and also by the increased efficiency of the system of collection.[7]

Year.				£
1632-3	...	...	...	22,553
1633-4	...	...	...	25,846
1634-5	...	...	...	38,174
1635-6	...	...	...	39,078
1636-7	...	...	...	38,889
1637-8	...	...	...	57,387
1638-9	...	...	...	55,582
1639-40	...	...	...	51,874

The restraints on trade engaged the attention of the Irish Parliament in 1640. One of the first bills transmitted

---

[1] Carte, vol. i, p. 169.   [2] *Ibidem.*   [3] Strafforde's *Letters,* vol. i., pp. 308, 380.
[4] C.S.P., Ire., 1633-47, p. xiii ; Cooper. *Life of Wentworth,* vol. i, pp. 128-30.
[5] C.S.P., Ire., 1633-47, p. 273.   [6] C.S.P., Ire., 1633-47, p. xxxii.
[7] *Contributions towards a History of Irish Commerce,* by Wm. Pinkerton, Ulster Journal of Archæology, vol. iii., p. 189.

to England under Poyning's Act was one "for the repeal of several penal laws prohibiting the exportation of linen and other commodities," and for leave to export freely without any licence to all parts of the world, corn, grain, malt, and, oatmeal.[1] The illegal raising of the Book of Rates was also complained of as causing "a general and apparent decay of trade."[2] In February, 1641, the House of Lords drew up articles of grievances complaining, amongst other things, that trade suffered by the heavy rates on commodities; of the proclamation for the sole emption and uttering of tobacco; the monopolies of starch, tobacco, pipes, soap, and glass; the seizing of linen yarn for not conforming to regulations; and the oppression and extortion of custom-house officers.[3] The House of Commons also complained of the same grievances. In reply to the grievances of the Lords the King consented that the Book of Rates should be reduced and the rate on goods settled at five per cent. *ad valorem*; that the duties on French and Spanish wines should be re-settled by Parliament so as not to injure the revenue; that the proclamation for the sole emption and uttering of tobacco should be revoked if Parliament would settle a rate of ninepence in the pound on that commodity; and that all monopolies except those of alnage and glass should be abolished.[4] In reply to the Commons the King consented to allow the transportation of wool without licence, provided it were to England; to abolish the licences for exporting sheepskins and pipestaves upon settling a small duty thereon; to have the duties on licences for the retailing of wine and aqua vitae settled by Parliament; and that wearing apparel and horses for private use should be imported and exported duty free; but refused to allow English money to be exported to Ireland, as this was a matter to be conceded by the English Parliament.[5] The Houses of Parliament were not altogether satisfied with these replies, and further negotiations took place.[6] It is possible that the majority of

[1] C.S.P., Ire., 1633-47, pp. 246-269. [2] *I.C.J.*, vol. i., p. 162. [3] Carte, vol. i, p. 245. [4] Carte, vol. i, p. 283. [5] Carte, vol. i, pp. 285-6. [6] Carte, vol. i, pp. 287-295.

Parliamentary demands might have been conceded, but all attention to matters of trade and commerce was rapidly dispelled by the outbreak of the Rebellion, which was itself stated to have been provoked in some measure by the restraints on the freedom of trade.[1]

The articles traded in did not change during this period, and consisted principally of wool, either licensed or smuggled, frieze, cattle, provisions, corn and fish. In 1618 the following were said to form the principal articles of trade :—" Live beeves, hides, tallow, corn, yarn, barrelled beef and fish, and from some of the Munster ports great proportion of woollen commodities such as caddowes, rugs, wool, mantles, Irish frieze and the like."[2] The imports, on the whole, exceeded the exports; in 1641, for example, the value of the duties on the imports was £20,125, and on the exports £14,470.[3] The following list[4] gives a good idea of what commodities were exported from Ireland in an ordinary year :—

YEAR ENDING, 20th MARCH, 1641.

Beeves	45,605	Train Oil, tun	...	96
Barrels of Beef	15,215	Linen Yarn	...	2,921
Butter, cwt.	34,817	Horses	...	199
Tallow, cwt.	20,136	Oats, qrs.	...	2,459
Candles, cwt.	617	Oatmeal, qrs.	...	492
Hides	134,121	Rapeseed, qrs.	...	1,036
Calf Skins, doz.	853	Rape Oil, tuns	...	63½
Sheep	34,845	,, tons	...	384
Sheepskins, cwt.	3,111	Boards and Planks	...	209
Lambskins	1,667	Barrel Staves	...	941
Wool, great stones	151,576	Hogshead Staves	...	663
Broadcloth, pieces	506	Pipe Staves	...	144
Frieze, yards	279,722	Goat Skins	...	360
Frieze Stockings, doz.		Kid Skins	...	231
pairs	4,287	Fox Cases	...	3,091
Rugs	4,778	Otter Skins	...	452
Caddowes and Blankets	6,589	Deer Skins	...	398
Herrings, barrels	28,311	Feathers, cwt.	...	230
Salmon, ton	526	Bacon, flitches	...	297
Pilchards, ton	1,263	Iron, tons	...	778
Hake, cwt.	830	Lead, tons	...	201

[1] *Desiderata Curiosa Hibernica*, vol. ii., p. 79.   [2] C.S.P., Ire., 1615-25, p. 184.
[3] C.S.P., Ire., 1633-47, p. 355.   [4] C.S.P., Ire., 1669-70, p. 55.

IN THE SEVENTEENTH CENTURY. 67

The period under consideration compared very favourably with later periods in one important respect, namely, the quantity of shipping owned by Irish merchants. "They have small ships in some sort armed to resist pirates for transporting commodities to France and Spain, yet no great number of them."[1] The existence of a fairly large amount of Irish shipping may be inferred from a complaint written in 1632 that the Dutch merchants were cutting the Irish out in the shipping business. "Before the Dutch traded here, there belonged above twenty ships and barks to the city of Dublin alone."[2] This decrease in the volume of shipping must have been but temporary, for we read that in the decade 1630-40 the shipping of Ireland increased a hundredfold,[3] a result which was achieved principally by the efforts of Wentworth to recover the Irish carrying trade from the Dutch.[4] Wexford was a place "mighty in ships and seamen."[5] As we shall see in a later section, the shipping of Ireland was entirely destroyed during the Rebellion, and was not replaced afterwards owing to the restrictions imposed by the Navigation Acts, and the destruction of the Irish woods.

We have confined all our observations on Irish trade to the foreign trade, and indeed there is very little to be said about the internal trade of the country. The vast majority of the inhabitants were practically self-supporting, and supplied all their own wants. The few commodities which they were forced to purchase, they obtained from "petty chapmen," or wandering pedlars, who were considered so necessary to the economy of the country that they were expressly excepted from the Vagrancy Acts.[6] The greater part of the towns were on the sea coast, and communication between them was made by sea, the means of internal communication being very primitive.[7] It was part of the plan of the English to build

[1] Fynes Moryson, *Itinerary*, p. 225; see *Advertisements for Ireland*, MS., T.C.D.
[2] Egmont MSS., p. 72.
[3] C.S.P., Ire., 1633-37, p. 252; Leland, *History of Ireland*, vol. iii., p. 41.
[4] Strafforde's *Letters*, vol. i., p. 104; vol. ii., pp. 434-5.
[5] *Ormond MSS.*, vol. ii., p. 185.   [6] C.S.P., Ire., 1662-6, p. 437.
[7] Bagwell, *Ireland Under the Tudors*, vol. iii., p. 442.

roads in Ireland with the object of facilitating the passage of troops;[1] and in 1614 provision was made for the maintenance and repair of all existing highways by the forced contribution of labour and materials by the inhabitants of each parish.[2] The repairing of bridges, causeways and toghers was further provided for twenty years later, when the imposition of a tax for this purpose in each county was authorised.[3] The roads, however, do not seem to have been satisfactory; Boate complained of the scarcity of bridges;[4] and in 1662 the " general defect in the highways of the Kingdom " engaged the attention of the House of Commons.[5] As late as the beginning of the eighteenth century the Irish roads continued to be wretched and frequently impassable.[6] Communication by means of the rivers was not widely practised. In 1615 the House of Commons petitioned for an Act for the clearing of certain rivers for the better transportation of merchandise, and the Government replied that this was a project which it was determined to encourage.[7] Nothing, however, was done. It was one of Wentworth's schemes to open up communication between the upper and lower Shannon by the removal of a rock which prevented it; he sent engineers to report on the matter; but again nothing was done. Boate complained that the other rivers of Ireland were impeded by cataracts, fords and weirs.[8] Communication was rendered still more uncomfortable and difficult by an almost total absence of accommodation for travellers.[9]

(10) *The Woollen Industry.*

In early times the Irish woollen industry received much encouragement from the Government of the country; indeed, the industries in England and Ireland were regarded with equal favour by the English Kings. By a Statute passed in 1338 all cloth-workers of whatso-

[1] C.S.P., Ire, 1600-1, p. 264. [2] 11, 12 & 13 Jac. I., c. 7.
[3] 10 Car. I., Sess. 2, c. 26. [4] *Natural History of Ireland*, 1652.
[5] *I.C.J.*, vol. i., p. 511; C.S.P., Dom., 1689-90., p. 365.
[6] O'Brien, *Economic History of Ireland in the Eighteenth Century*, p. 360.
[7] *I.C.J.*, vol. i., p. 34. [8] *Ireland's Natural History*, 1652.
[9] *Advertisements for Ireland*, MS., T.C.D.

ever country were invited into Ireland as well as into England, Scotland and Wales. Many other Acts were passed by the English Parliament to encourage the manufacture in England, and these Acts all became law in Ireland in 1495 by virtue of Poyning's Act, and were unrepealed at the time of the Restoration.¹ It was with the object of encouraging the manufacture of Irish woollens that the export of raw wool was prohibited in 1522.² In 1569 the total prohibition imposed by the last-mentioned Act was replaced by very heavy customs duties on the export of wool, flock, woollen yarn, and sheepfells,³ and two years later the evasion of these customs duties was constituted a felony.⁴ At the same time a restraint was put on the exportation of the finished product by the enactment that no cloth or other work or stuff made or wrought of wool should be exported except by merchants within the staple cities and towns, or by free merchants of the boroughs and privileged towns.⁵ This Act contained a special provision that English merchants might take Irish cloth in exchange for their own wares, outside the staple towns. This policy was dictated by the fear that if Irish cloth were permitted free access to Flanders it might undersell English cloth, and that, if the Flemish merchants obtained Irish wool, they might develop their own industry.⁶ Evidence of this policy is to be found in the charter to the city of Galway, which forbade the exportation of woollens from there, Galway being one of the ports from which Irish commodities were shipped to foreign countries, but not to England.⁷ In spite of the statutory prohibitions, licenses were frequently given to English dealers to export wool from Ireland.⁸ Lord Deputy Sidney attempted to stop this evil by proposing an Act restraining the Lord Deputies from granting licenses for the export of wool, but the interests of the English dealers again prevailed.⁹ Lord Deputy Fitzwilliam went

¹ C.S.P., Ire., 1666-9., p. 13. ² 13 Henry VIII., c. 2. ³ 11 Eliz., Sess. 3, c. 10.
⁴ 13 Eliz., c. 2. ⁵ 13 Eliz., c. 1.
⁶ Green, *Making of Ireland and its Undoing*, p. 145.
⁷ Hardiman, *History of Galway*, p. 82. ⁸ Green, *op. cit.*, p. 146.
⁹ *Ibidem*, pp. 146-8.

a step further by endeavouring to gain the profit of the export of wool to England for his own pocket, but the Queen had already granted a license to export large quantities to an English merchant.[1]

As a result of Elizabeth's policy, the Irish clothing industry declined, and the English industry correspondingly flourished. " With the decline of manufactures in their own land, Dublin merchants made suits for free import of English cloth. The complaints of the towns and of their burghers forced to take to tillage tell, too, of the destruction of manufactures and of the old exchange of trade. Elizabeth did not die till her work of destruction was done. A few years later it could be urged in the English Parliament, as an argument for allowing Irish cattle to be sold in England, that half the money received for them by the Irish was spent in England, as ' there is no household stuff used in Ireland but what comes out of England.' "[2] In 1607 " merchandize for apparel " is mentioned as a commodity imported from England.[3]

However, the English hostility had not completely ruined the Irish woollen trade even at the beginning of the seventeenth century. The rigid blockade of the Irish ports was sometimes successfully evaded, and a great part of the munitions used by the Irish in their wars against Elizabeth was brought from Spain in exchange for wool.[4] Moreover, the people of Ireland themselves wore woollen garments universally. " In Elizabeth's reign the long woollen mantle was the habitual covering of the Irish, both men and women; underneath was generally worn a linen smock or shirt. Later on, the men wore breeches and short coats made of coarse cloth of different colours, coarse woollen stockings, woollen caps or hats, and over all the same a large mantle of frieze. The number of woollen garments worn by women and children also

---

[1] *Ibidem*, p 178; and see generally *Contributions Towards a History of Irish Commerce*, by Wm. Pinkerton, Ulster Arch. Journal, vol. iii, pp. 181-2; Cunningham, *English Industry and Commerce*, vol. ii., p. 369; Otway's *Report on the Handloom Weavers*, 1840.
[2] Green, *op. cit.*, p. 152. [3] C.S.P., Ire., 1606-8, p. 81. [4] C.S.P., Ire., 1601-3, p. 251.

increased, and altogether the amount of wool required to make the garments of one individual must have been considerable. This native cloth was, no doubt, very coarse and rough. It was made by the women in their own homes and the greater part of it must have been used by the people themselves."[1]

The industry, then, at the beginning of the century seems to have been practically restricted to the supplying of this demand for rough articles of clothing for home consumption, and the manufacture of woollen goods for export must have almost totally ceased.[2] "They work not their woollens here," wrote Lord Chichester in 1614,[3] and in the following year it was suggested in the House of Commons that clothworkers should be sent from England to relay the foundations of a woollen manufacture.[4] In the absence of such a manufacture, the restrictions on the export of raw wool pressed as an intolerable grievance upon the numerous sheep farmers of Ireland, and the House of Commons petitioned for a relaxation of these restrictions. The English Government, however, refused to grant this concession,[5] and when a bill to effect this object was transmitted to England under the provisions of Poyning's Act, it was not returned.[6]

But although the English Government objected to the export of Irish wool, which might enable Spanish and French clothiers to injure the English woollen industry, it had no objection to the establishment of an industry to work the wool up in Ireland, and to attain this object the project of erecting a new system of staple towns in Ireland was adopted. The double advantage to be gained by such a system was that the producers of wool would be assured of commanding a large market and a steady price for whatever quantities of wool they wished to sell, and that English clothworkers would be encouraged to settle in Ireland owing to the restraints on the exportation of

---
[1] Murray, *Commercial Relations*, p. 96.
[2] In the list of apparel appointed to the soldiers in the Derbyshire levy in 1600 was "one pair of Kersey stockings or two pairs of Irish frieze . . . and an Irish mantle." *Rutland MSS.*, vol. i., p. 357.   [3] C.S.P., Ire, 1611-14., p. 506.
[4] *I.C.J.*, vol. i., p. 52.   [5] *I.C.J.*, vol. i, p. 34.   [6] *I.C.J.*, vol. i, p. 56.

the raw material. In the course of the years 1616-7 the new staples were erected, and Dublin, Drogheda, Cork, Limerick, Galway, Carrickfergus, Londonderry and Youghal were declared staple towns.[1]

How far the staples succeeded in attaining the objects aimed at in their erection is a matter of some doubt, and is clouded by contradictory evidence. On the one hand, the Lord President and Council of Munster wrote to the Lord Deputy representing " the many evils arising from the new staples." They pointed out that, instead of English clothworkers being attracted from England to set up a manufacture, the factor of the staples engrossed all the wool and shipped it to England; that a great number of English and Irish already employed in making frieze, mantles, caddowes (coarse blankets), and other commodities had consequently been thrown out of employment; that the sheep-farmers were giving up the rearing of sheep and were subletting their lands to the Irish " to be tilled at the fourth sheaf," and that the smuggling of wool to Spain was greatly on the increase.[2] To these complaints the factors of the staples replied that the erection of the staples had had the effect of hindering the exportation of wool; that the wool came to be worked instead of exported; and that some Englishmen " have embarked large capitals in the manufacture, and have made some quantity of cloth with which they have furnished the Lord Deputy himself, but the works were interrupted by complaints and oppositions."[3] Again, in 1622, we hear of a " goodly clothing work of Sir T. Roper near Dublin wherein at carding, spinning, weaving, working, dressing, and dyeing cloth, many poor people are daily set on work."[4] and in a manuscript written in that year it is stated that the customs duty on the export of rugs and wool coverlets was considerable.[5] Whatever progress the manufacture

[1] C.S.P., Carew, 1603-24, pp. 320, 334; C.S.P., Ire., 1615-25., pp. 150, 292.
[2] C.S.P., Ire., 1615-25, p. 252; C.S.P., Carew, 1603-24, p. 424.
[3] C.S.P., Ire., 1615-25, p. 274.   [4] C.S.P., Ire., 1615-25, p. 361.
[5] *Advertisements for Ireland*, MS., T.C.D.

made in Munster was probably due less to the staples than to the encouragement of the Earl of Cork.¹

However, although the staple system may have had the effect of encouraging the industry in some slight degree, it certainly did not succeed in putting an end to the evils which it was meant to abolish, for in 1623 the question of the woollen manufacture again engaged the serious consideration of the Commissioners of Trade. This body sensibly recommended that one of two courses should be adopted; either the production of the raw material should be encouraged, in which case the restraints on exportation must be relaxed; or the working up of the raw material in Ireland should be aimed at, in which case there should be no restraint, as there then was, on the exportation of fuller's earth and wood ashes from England to Ireland. They also recommended the adoption of a non-importation agreement.²

These suggestions, however, were not adopted. The transportation of fuller's earth continued to be prohibited³ and the restrictions on the exportation of wool were rigidly enforced. In 1623 the export, except to England and Wales, was restrained more strictly than ever,⁴ and five years later no relaxation of this order had been made.⁵ Three years later even this limited market was partially closed, for it was provided that no wool should be transported to England without a special warrant from the Lords Justices, a regulation which, it was foreseen at the time, " would breed great discontent."⁶ However, the difficulty was to some extent surmounted by the smuggling of large quantities of wool abroad.⁷ Great quantities of Irish wool were annually shipped to England. " The town of Norwich had bought in one year some years since a hundred thousand pounds worth of Irish yarn, and I am assured never receives more yarn than Bristol, Manchester, and other towns here from

---

¹ Townshend, *Life and Letters of the Great Earl of Cork*, p. 80 ; on the question of the merits of the staple system see Cunningham. *Growth of English Industry and Commerce*, vol. ii, p. 298.   ² C.S P., Ire., 1615-25, p. 424.
³ This appears from the State paper last quoted.   ⁴ C.S.P., Ire., 1615-25, p. 402.
⁵ C.S.P., Ire., 1625-32. p. 331.   ⁶ C.S.P., Ire., 1625-32, p. 616.
⁷ Strafforde's *Letters*, vol. i., p. 423 ; *Advertisements for Ireland*, MS., T.C.D

thence."[1] Great quantities were also smuggled to the Continent. "Now I understand that the yarn of the provinces of Munster and Connaught is transported to Spain, the Low Countries and France, where it is very well bought."[2]

When Wentworth came to Ireland in 1633 he found "very little trade amongst the Irish, scarce any manufacture at all, only some small beginnings towards a clothing trade, which was likely to increase in time. This he thought proper to discourage all he could, as apprehending that it would in the end be an infinite detriment to the clothing, the staple commodity of England. He considered, further, that, in reason of state, so long as the Irish did not manufacture their own wool, they must of necessity fetch their clothing from England, and consequently in a sort depend upon it for their livelihood, and be disabled to cast off that dependence without nakedness to themselves and children."[3] "To serve your Majesty completely well in Ireland," he wrote, "we must make sure still to hold them dependent upon the Crown, and not able to subsist without us, which will be effected by wholly laying aside the manufacture of wools into cloth or stuff there, and by furnishing them from this Kingdom."[4] Wentworth's advice was acted on in England, when in 1636 the Irish Committee of the Privy Council resolved that great care must be taken that no fuller's earth was allowed to be sent from England to Ireland lest the Irish clothing manufacture might increase.[5] Wentworth also encouraged the granting of licenses for the transportation of wool to England. "I writ to you for a dormant warrant to give licenses for the transportation of wool hence into England, which must of necessity be allowed, both in reasons of state, for the bettering of his Majesty's customs, and because they have no means here to manufacture it themselves, so as the commodity would be entirely lost unto the growers, until this expe-

[1] *Advertisements for Ireland*, MS., T.C.D. [2] *Ibidem*.
[3] Carte, vol. i., p. 170; Strafforde's *Letters*, vol. ii, p. 19.
[4] Strafforde's *Letters*, vol. i., p. 93. [5] C.S.P., Ire., 1633-47, p. 136.

dient be granted; and to entrust the exportation into the hands of the merchant staplers is a remedy worse than the disease; besides their charter is forfeited through insurance, and shall never be renewed unto them again by my advice, it being a corporation newly set up by the device of my lord Grandison, only as they say here to put some crowns into his purse."[1]

The Irish Parliament of 1640 paid much attention to the difficulties under which the woollen industry laboured. In 1641 a Bill to repeal the restrictive statute of Elizabeth was read in the House of Commons, but never passed into law.[2] These restrictions were always enumerated amongst the grievances for the redress of which Parliament petitioned the King.[3] As a result of these petitions permission was given to export wool freely to England, but not to foreign countries.[4]

The volume of the export trade in wool and woollens may be judged from the following figures:—[5]

EXPORTS FROM IRELAND, Year Ending 25th March, 1641.

Wool, great stones	151,576
Broadcloth, pieces of 36 yards	506
Frieze, yards	279,722
Frieze Stockings, dozen pairs	4,287
Rugs	4,778
Caddowes and Blankets	6,589

There is no doubt that, although the proportions the woollen industry attained in this period were negligible compared with those it reached at a later date, it was the great staple manufacture of the early seventeenth century. "What they had of trade was chiefly in the hands of the Irish, who had little manufacture but rugs, mantles, and friezes."[6]

(11) *The Linen Industry.*

It has frequently been asserted that the Irish linen industry was founded by Wentworth, but recent research

[1] Strafforde's *Letters*, vol. i, p. 202. [2] *I.C.J.*, vol. i., p. 164.
[3] C.S.P., Ire., 1633-47, p. 294. [4] C.S.P., Ire., 1633-47, p. 329.
[5] C.S.P., Ire., 1669-70, p. 54.
[6] Laurence, *The Interest of Ireland in its Trade and Wealth Stated*, London, 1682, p.4.

has shown that this statement is based on a complete ignorance of early Irish industrial conditions. The truth is, that, until the Elizabethan war upon Irish trade, the linen industry in Ireland occupied at least as important a place and was as fully developed as the woollen industry. The fiction that Ulster owes its great prosperity to Wentworth, who laid the foundations of a great linen industry to compensate Ireland for the woollen industry, which he proposed to discourage, is a convenient one for English historians who wish to present in as favourable a light as possible the less justifiable measures of English policy in Irish affairs. It is also asserted from time to that that even if Wentworth were not to be thanked for the growth of the modern linen industry, some credit must at least be given to the English Parliament, which, in suppressing the Irish woollen industry in 1698, expressed a pious hope that the linen manufacture might flourish in its place. As a matter of fact the British Parliament afterwards did all it could to impede the progress of the Irish linen industry, and whatever success it achieved was due to the encouragement not of the English but of the Irish Parliament.[1]

The question of the condition of the linen manufacture in Ireland before the sixteenth century is fully dealt with by Mrs. Stopford Green. "Linen was sold on the stalls of every Irish market, and was carried abroad; and flax was grown in every part of Ireland from North to South. Foreign writers attest the great abundance of linen in Ireland. 'Ireland,' they say, 'abounds in lint which the natives spin into thread, and export in enormous quantities to foreign nations. In former ages they manufactured very extensively linen cloths, the greater portion of which was absorbed by the home consumption, as the natives allowed thirty or more yards for a single cloak, which was wound or tied up in flowing folds. The sleeves also were very capacious,

---

[1] O'Brien, *Economic History of Ireland in the Eighteenth Century*, pp. 194-6; Cunningham, *Growth of English Industry and Commerce*, vol. ii, p. 522.

extending down to the knees. But these had nearly gone out of fashion in 1566."[1] Need I mention the common linen covering which the women wore in several wreaths on their heads, or the hoods used by others; for a woman was never seen without either the veil or a hood on her head, except the unmarried, whose long ringlets were tastefully bound up in knots, or wreathed around the head and interwoven with some bright-coloured ribband. If to these we add the linens for the altar, the cloths for the table, the various linen robes of the priests, and the shrouds which were wrapped around the dead, there must have been a great abundance of linen in Ireland. . . . . Charged by the English with, amongst other vices, an extravagant use of linen in their dress, they nevertheless provided for foreign markets, exporting linen cloth and fadings to Chester, the Netherlands, and Italy."[2] The presence of the linen industry in Ireland on a large scale as late as the middle of the sixteenth century is attested by the Statute of 1539, which forbade more than seven yards of linen to be used in any shirt,[3] and by that of 1569 forbidding the watering of hemp and flax in running streams.[4]

However, the linen industry suffered with every other Irish industry during the Elizabethan wars; the condition of the country was in itself enough to unsettle all the occupations of peace; Irish trade on the high seas was harassed by English pirates; and the linen industry must have suffered especially from the illegal but officially encouraged export of linen yarn in large quantities to England.[5] At the beginning of the seventeenth century the industry was probably at a very low ebb, if not practically non-existent. In 1603 it was suggested that flax should be sown in great quantities in Ireland, as in a short time the Irish linen trade, if encouraged, would rival those of Spain and the Low Countries;[6] and the

[1] *Cambrensis Eversus*, vol. ii., p. 169.
[2] Green *The Making of Ireland and its Undoing*, pp. 49-50.
[3] 33 Henry VIII., c. 2.   [4] 11 Eliz., c. 5.
[5] Green, *The Making of Ireland and its Undoing*, pp. 145-52.
[6] *Memorials for the Better Reformation of the Kingdom of Ireland*, C.S.P., Ire., 1603-6, p. 134.

sowing of a certain amount of flax was made one of the conditions on which lands were granted to planters in Leitrim.¹ This would seem to show that there was at that time but a small manufacture in existence, but that the progress and prospects of the linen industry of fifty years before was still remembered. One of the motives offered " to induce the City of London to undertake the Plantation of the North of Ireland " was that Ulster would furnish the requisites for " thread, linen cloth, and stuffs made of linen yarn which is finer there and more plentiful than in all the rest of the Kingdom."² Amongst the Acts sent from England in 1611 as being " thought fit to be propounded at the next Parliament in Ireland " was one entitled " an Act for sowing hemp and flax and for making linen cloth."³ This Act never passed into law; and the suggestion was renewed in 1623 by the Committee of Trade.⁴

It is doubtful how far the industry progressed during these years, but it certainly did not attain the point where Ireland was able to use all her own yarn, for we hear of licenses being granted to export large quantities of linen yarn from time to time.⁵ The granting of such licenses was refused by James I. during the earlier years of his reign in order to encourage the industry, but when it became apparent that the industry was not growing, the practice of granting licenses was resumed.⁶ A certain amount of linen must, however, have been worked up in Ireland to meet the home demand; the women wore head-dresses made of " rolls containing twenty bands of fine linen cloth . . . . and a muffler of even quantity."⁷ One of the most flourishing centres of the manufacture at this time was at Bandon,⁸ and it was also set up at Youghal by the Earl of Cork.⁹

When Wentworth came to Ireland he was faced with a difficult problem. On the one hand he was anxious to

---
¹ *Desiderata Curiosa Hibernica*, vol. ii, p. 63. ² C.S.P., Ire., 1608-10, p. 208.
³ C.S.P., Ire., 1611-14, p. 250. ⁴ C.S.P., Ire., 1615-25, p. 426.
⁵ C.S.P., Ire., 1611-14, p. 379. ⁶ Rep. Pat. Rolls, James I, vol. i, part i, p. 102.
⁷ Gernon, *Discourse of Ireland*, 1620. ⁸ *Cork Arch. Jnl.*, vol. iii, o.s., p. 285.
⁹ Townshend, *Life and Letters of the Great Earl of Cork*, p. 100.

increase the wealth and revenue of Ireland for the sake of the King, while on the other hand he felt bound to do all he could to discourage the largest manufacture of the country, lest it should injure the English weavers. His way out of the difficulty was to encourage the linen industry at the expense of the woollen. " He observed that the soil of Ireland was very fit for bearing of flax ; and that the women were all naturally bred to spinning ; and therefore resolved to put them upon making of linen cloths, which he thought would be rather a benefit than a disadvantage to England; and as the way of living and consequently the labour of the Irish was exceeding cheap, he did not question but they might soon be enabled to undersell the linen cloths of Holland and France by at least twenty in the hundred. With this view he sent to Holland for flax seed (it being of a better sort than any they then had in Ireland), and into the Low Countries and France for workmen. The flax was sown and took very well with the soil; looms were set agoing and people employed; provision was made by fitting regulations about their yarn and cloth to ensure them both to be of a proper goodness, upon which the sale of any commodity depends. . . . . . He gave an uncommon instance of his public spirit on this occasion ; and as all projects are in the beginning attended with great expenses, so that they who first set about them are often ruined, when their successors get estates; so to encourage others, he engaged in it himself, venturing his own private fortune and spending thirty thousand pounds in a work, which nothing could have moved him to undertake but a laudable zeal for the good of the Kingdom which he governed."[1]

The efforts of Wentworth met with some success, as both the quantity and quality of Irish yarn and cloth improved in the next few years.[2] Wentworth stated that his Irish looms could undersell those of France and Holland by

---
[1] Carte, vol. i, pp. 170-1 ; Strafforde's *Letters*, vol. i, p. 473.
[2] Murray, *Commercial Relations*, p. 113; Smith, *Irish Linen Trade Handbook*, Belfast, 1876, p. 29; Temple, *Essay on Trade of Ireland*, 1673.

80   THE ECONOMIC HISTORY OF IRELAND

twenty per cent.¹ The extent to which the industry grew in a short time may be judged of by the fact that among the grievances complained of by the Irish House of Lords in 1641, one was that "many thousand hundredweight of linen yarn and great quantities of linen cloth have been confiscated by force from poor people for want of breadth and a proper number of threads,"² but the King refused to assent to the repeal of the regulations in question, as this would "have been the ruin of that trade."³ In the year ending 26th March, 1641, 2,297 hundredweight of linen yarn was exported, but no linen cloth seems to have been exported that year.⁴ Unfortunately, the Rebellion broke out in that year, and in the terrible decade that followed the growing industry was completely ruined.⁵

(12) *The Fishing Industry.*

The greater part of the fishing carried on in the seas adjoining Ireland was in the hands of foreigners, with whom the Irish made little or no attempt to compete. From a very early time foreign fishermen had been accustomed to fish in Irish waters; an Act of the reign of Edward IV. provided that all foreign vessels coming to Ireland to fish were to pay a duty to the Crown, and the same Statute made provisions for the careful packing and merchandizing of Irish fish.⁶ This Act ceased to be enforced during the disturbed years of Elizabeth's reign, but it was again put into operation in 1607.⁷ In 1615 the Irish Government attempted to levy this tax on English vessels engaged in the Irish fisheries, but the attempt was rendered unsuccessful by the action of the English Government.⁸ In spite of this tax the greater part of the profits of the rich fisheries in Irish waters continued to be gained by foreigners,⁹ who sometimes succeeded in

---
¹ Cunningham, *Growth of English Trade and Commerce*, vol. ii. p. 369.
² C.S P., Ire., 1633-47, p. 262.   ³ Carte, vol. i. p. 284.   ⁴ C.S.P., Ire., 1669-70, p. 54.
⁵ Murray, *Commercial Relations*, p. 114 ; Smith, *Irish Linen Trade Handbook*, p. 30.
⁶ 5 Ed. IV., c. 6.
⁷ C.S.P., Ire., 1606-8 p. 234. A lease was made of the fines arising from this Act in 1612, *Harris MSS.*, National Library of Ireland, vol. vi., p. 136
⁸ C.S.P., Ire., 1615-25, p. 59.   ⁹ C.S.P., Ire., 1615-25, p. 426 ; C.S.P., Ire., 1633-47, p. 68.

avoiding payment of the duty.¹ "I have heard some of the Irish merchants affirm that, were they strong in ships, and able to undergo completely voyages into foreign lands, they might transport from Ireland 20,000 tuns of pilchards, as much in herrings some years, and near the same quantity of salmon, and likewise 30,000 tons of dry fish; and most of this is now enjoyed by foreigners, who fish upon these coasts and carry all away, and pay no custom to the King for the same. They are seen to make up upon the Irish coast about the number of five or six hundred sail yearly, and all deep laden with fish."² Various attempts were made by the Irish Government to encourage the fisheries. In 1615 a bill was sent to England proposing to restrain all foreigners from fishing on the Irish coast, but the bill was not returned "for reasons of state."³ Again, in 1623, the Commissioners of Ireland reported that "as the fishing in general of Ireland was for the most part in the hands of strangers, no foreigners should be permitted to fish upon these coasts."⁴ The Irish House of Commons, in 1634, passed a bill for the regulation of the herring fishery, but this bill did not receive the royal assent.⁵

There were also attempts made by private enterprise to improve the fisheries, but they do not seem to have come to anything. In 1623 forty gentlemen applied for a patent "with a view to overthrow the gain of the Hollanders by fishing, and to set twenty thousand people on work." They set out the great possibilities which an Irish fishing industry possessed:—"The Northern seas afford the Hollanders but two sorts of fish, cod and herrings. Ireland yields a great plenty of both, besides these several fishes following: pilchards, the best in Europe, which are vented in the Straits; hake and cod much esteemed in Biscay; ray and conger for Brittany; salmon and buckthorn desired in all countries; cod and ling with the train oil that comes of them

¹ C.S.P., Ire., 1615-25, p. 428.   ² *Advertisements for Ireland*, MS., T.C.D.
³ C.S.P., Ire., 1615-25, p. 27.   ⁴ C.S.P., Ire., 1615-25, p. 426.
⁵ I.C.J., vol. i., pp. 85-87.

vented in England."¹ At the same time an effort was made to establishing fishing off the coasts of Mayo,² and a determined attempt to improve the Munster fishing was made by the Earl of Cork, who built boats and established salting and fish houses.³

One of the chief difficulties in the way of a successful fishery was the supply of salt. For instance, in 1628, La Rochelle was besieged, and many fishermen were ruined in consequence, as they could get no salt.⁴ The fishing boats were also troubled by pirates. In 1632 the pilchard fishery off the south coast was stopped by the ravages of the Turks, and a considerable loss was caused to the coast towns, " where frequently twenty thousand pounds a year came by reason of the fishery."⁵ The sea, however, was successfully cleared of pirates by Wentworth, and two years later we read that " nothing is heard of the Turks, and the take of pilchards is the largest on record."⁶

In 1641 there were complaints of the failure of the pilchard fishery,⁷ and the total exports of fish in that year were as follows :—⁸

Herrings, barrels	23,311
Salmon	526
Pilchards, tons	1,263
Hake, cwt.	830
Train Oil, tons	96

We read in a manuscript written in 1622⁹ of the great trade in ambergrease that was at that time carried on. " Much ambergrease is had and gathered on the South and West Coasts. . . . The Galway merchants bring a great quantity yearly into England and Spain of this ambergrease, which is held to be as good and perfect as any."

¹ C.S.P., Ire., 1615-25, pp. 403, 580. ² C.S.P., Ire., 1615-25, p. 404.
³ Townshend, *Life and Letters of the Great Earl of Cork*, p. 101-2.
⁴ C.S.P., Ire., 1617-60, p. 145 ; Strafforde's *Letters*, vol. i., p. 96,
⁵ C.S.P., Ire., 1625-32, p. 645. ⁶ C.S.P., Ire., 1633-47, p. 68.
⁷ C.S.P., Ire., 1633-47, p. 255. ⁸ C.S.P., Ire., 1669-70, p. 54.
⁹ *Advertisements for Ireland*, MS., T.C.D.

(13) *Minor Industries.*

There are some traces of the existence of a glass manufacture on a small scale. In 1608 a patent was granted to one Adam Whitty, of Arklow, for the making of glass for ten years; in 1618 a Venetian merchant was anxious to set up a glass industry in Munster;[1] there are still extant the accounts of glassworks set up at Ballynegery in 1622;[2] and Sir Abraham Bigs carried on glassworks at Birr from 1623.[3] In 1634 the glass industry must have been extensive, for "clay for making glass" was one of the principal commodities imported from Spain.[4] In 1634 Sir Percival Hart was granted a monopoly of making in Ireland "black glass drinking cups and vessels and pots similar to those made at Murano in Italy and commonly called Venice drinking glasses"; a privilege which was to last for twenty-one years in consideration of an annual payment of fifty marks.[5] Boate said that "numerous glass houses had been set up in Ireland, none in Dublin or other cities, but all of them in the country; amongst which the principal was that of Birr."[6]

There is no evidence that any manufacture of pottery was carried on in the first half of the seventeenth century except the statement of Boate that "in every part of Ireland there is found a kind of clay very fit to make bricks and all sorts of pottery ware . . . . a great number of English potters in several parts of the land have set up their trade."[7]

Beer made of hops was not made in this period in Ireland, but there was an extensive brewing of strong ale, great quantities of which were consumed. "I am not to speak of a certain kind of commodity that outstretcheth all that I have hitherto spoken of, and that is the selling of ale in Dublin; a quotidian commodity that hath vent in every house in the town every day in the week, and every hour in the day, and every minute in the hour.

[1] Townshend, *Life and Letters of the Great Earl of Cork*, p. 106.
[2] Marsh's Library MSS. vol. 23. 1, 3, No. 12.
[3] Westropp, *Glassmaking in Ireland*, Proc. R.I.A., vol. xxix., c. p. 34.
[4] Strafforde's *Letters*, vol. i., p. 104.   [5] C.S.P., Ire., 1633-47, p. 37.
[6] *Ireland's Natural History*, ch. 20.   [7] *Ireland's Natural History*, ch. 20.

There is no merchandise so vendible, and is the very marrow of the commonwealth in Dublin."[1] Wentworth was of opinion that a very considerable addition to the revenue could be obtained by means of a tax on brewing.[2] English taste disapproved of the Irish ale. "Scarce anywhere out of Dublin and some few other towns will you meet with any good beer or any seasonable bread for your money, only you may have some raw, muddy, wholesome ale, made solely of oats, which they buy there for fivepence the quarter, and commonly for fourpence, and yet they sell their ale pots dearer than in England they do the best beer. Now, if barley were grown there in plenty, and likewise in head towns, parishes, and thoroughfares and villages of note, malt houses and common brew houses were erected, and men experienced in those trades from England were employed, it would prevent this general mischief, which, if any weighty exigent would call thither any number of British men, would soon cause them to perish by their poisonous drink."[3] Much usquebaugh was distilled in all parts of the country,[4] and usquebaugh frequently figures among the lists of Irish exports of the time.

There was one industry which should have been developed as a result of the growing trade in provisions, namely, the tanning industry, but, unfortunately, most of the hides produced in Ireland were exported in their raw condition. For example, in the years 1605-8 there were exported from Dublin 6,100 raw and only 1,510 tanned hides.[5] The tanning industry was hindered from growing to some extent by the destruction of the woods and the consequent shortage of bark;[6] and partly by the law which prohibited the setting up of a tannery without a license.[7] The industry was further hampered by the engrossing of the raw hides by English merchants:—
"There would be tanners in and adjoining market towns if hides were brought for them to buy, but being fore-

[1] Fynes Moryson, *Itinerary*; Bagwell, *Ireland Under the Tudors*, vol. iii, p. 449.
[2] Strafforde's *Letters*, vol. i., p. 192. [3] *Advertisements for Ireland*, MS., T.C.D.
[4] *Ulster Jl. of Arch.*, vol. vii., o.s. p. 40. [5] C.S.P., Ire., 1608-10, p. 198.
[6] C.S.P., Ire., 1625-32, p. 331. [7] 11 Eliz., c. 3.

stalled by the merchants the tanners want employment, and thereby arises defect in one of the wealthiest trades."¹ As it was, there was not a tanner within twenty miles of most market towns, and but one skilful tanner in the whole of Ulster,² a defect which the planters at Derry and Coleraine sought to remedy.³ The old condition of affairs, however, continued, in spite of Wentworth's attempt to repeal all restrictions on the liberty of tanning,⁴ and in the year immediately before the Rebellion, 134,121 raw hides were exported from Ireland, whereas no tanned hides figure in the list of exports.⁵ Even after the outbreak of the Rebellion, when all other Irish trade had ceased, there still continued an export of raw hides.⁶

There is some evidence that shipbuilding was carried on on a small scale during this period. In the early years of the century the East India Company erected ironworks and built ships in Munster, but, if we are to believe contemporary documents, they were harassed by the "wylde Irish." Ships of four and five hundred tons were launched at these works, but the whole enterprise came to an end in 1613 through the hostility of the inhabitants of the neighbourhood.⁷ Ships, probably small, were also constructed in Mayo in connection with the schemes for developing the Irish fisheries.⁸

(14) *Public Finance.*

The character of the Irish revenue was fundamentally changed after the Restoration, when it assumed the form which it maintained in the main until the Union. In the reigns of the two first Stuarts the old English ideas on the subject still prevailed ; the King was expected, as far as possible, to " live on his own " ; and recourse to Parliament for additions to the revenue was the exception rather than the rule. Indeed, the whole constitutional struggle of the early half of the seventeenth century

---

¹ C.S.P., Carew, 1603-24, p. 206.   ² C.S.P., Carew, 1603-24, p. 204.
³ C.S.P., Ire., 1611-14, p. 225.   ⁴ Strafforde's *Letters*, vol. i, p. 315.
⁵ C.S.P., Ire., 1669-70, p. 54.   ⁶ *Portland MSS.*, vol. i, p. 78.
⁷ C.S.P., Ire., 1611-14, pp. 369-381.   ⁸ C.S.P., Ire., 1615-25, p. 404.

arose from the attempts of the Kings to supplement their income by use of their prerogative; just as the struggle of the present century between the House of Lords and House of Commons turned on the question whether the former had any right to interfere in the matter of supply; and the ultimate victory of the Parliament was really an assertion of the principle that the taxation of the subject is a matter to be decided only by the popular assembly. This struggle between the Stuarts and their Parliaments is not of mere constitutional interest. On the contrary, it was a powerful factor in the formation of English policy towards Irish prosperity. As long as the King possessed a large revenue of his own, absolutely beyond the control of Parliament, it was obviously his interest to augment as much as possible the sources from which that revenue was derived. To the King, moreover, it was a matter of indifference whether that revenue came from England or Ireland as long as it came into his Exchequer. It was, therefore, the interest of the English Government of that date to increase the wealth and trade of England and Ireland indiscriminately, as increased prosperity meant an increased revenue. And so we find, in fact, that, practically speaking, up to the period of the Restoration, the Kings of England were equally anxious to advance the wealth of their two Kingdoms. The war on Irish trade which we have described as taking place in Queen Elizabeth's reign was an exception, not real but apparent, as the trade war was not waged with the object of diminishing the wealth or trade of Ireland absolutely, but of transferring that trade from Irish to English hands, or, in other words, of transferring it from the possession of those who resented the payment of taxes to the English Government, to the possession of those who would pay such taxes willingly. This was a totally different and economically much less noxious policy than that which, at a later date, aimed at the destruction of Irish wealth and trade as a whole, a policy which was dictated very largely by the fear of the English Parliament that a prosperous Ireland would yield such large revenues to the English King as to place him outside Parliamentary control. This change of policy was the direct result of the change which took place in the seventeenth century in the power over the revenue.

The revenue of Ireland under James I. and Charles I. was divided into two kinds, the ordinary and extraordinary, the former being the absolute hereditary property of the sovereign accruing to him by virtue of his prerogative, and the latter consisting of whatever additions might be made to the ordinary revenue on special occasions by Parliamentary subsidies or other means. The ordinary revenue was itself divided into certain and casual. The certain ordinary revenue consisted of the Crown rents arising from the confiscations made by Henry VIII., Elizabeth, and James I., the composition rents which had been reserved to the Crown in 1585 in the Composition of Connacht, and the fees paid for licenses for the exportation of wool, yarn, and other prohibited commodities. The casual ordinary revenue was made up of a variety of smaller duties, the profits from the Court of Wards, respite of homage and other feudal dues, the import of wines, fees payable for seals and writs, forfeiture of felons' goods, the green wax money, the Star Chamber fines, twentieth parts, licenses for the sale of spirits, and other small miscellaneous duties. The customs were also classed amongst the casual revenue, but they were founded on an Act of Parliament whereby twelve pence in the pound was levied on the value of all goods, except wine and oil, imported to and exported from Ireland.[1] The King was also entitled at common law to an import and export duty of three pence in the pound on the goods of strangers.[2]

When this ordinary revenue proved insufficient to meet the expenses of the Government, it was augmented in various ways. During the later years of Elizabeth and the earlier years of James I. the English Government

---

[1] 15 Henry VII.   [2] C.S.P., Ire., 1611-4, p. 194.

was at great expense in maintaining an army originally of conquest, and subsequently of occupation, to subdue Ireland, and one of the consequences of that army's operations was to impoverish the country to such an extent that it was quite unable to furnish a revenue sufficient to pay the cost of its own subjugation. In this case the deficit in Ireland was made good by money sent out of England. It was the memory of this expenditure that caused so many English writers to complain of the uselessness of Ireland, which had cost England more than it ever could repay.

This was, however, a very exceptional means of finding money to pay the expenses of Irish Government, and, as soon as the country settled down, efforts were made to extract from the Irish taxpayer whatever additional revenue was required. The Irish Parliament, therefore, granted several subsidies to be levied on the lands and goods of the nobility, clergy and commonalty. These will be more fully described below.

A third method resorted to by government in need of money was to bargain with certain classes of the community to pay for some concession—or more commonly for the promise of some concession. Of this nature was the bargain effected by Charles I. with the Irish gentry in 1628—the famous Graces by which, in return for certain promises on the King's part, payments of £20,000 a year for five years were secured.[1] Attempts were also made to raise money simply by proclamation, without any statute. For example, in 1627, Charles I. increased the duties payable on the export from Ireland of many kinds of commodities,[2] and in the following year direct taxes were imposed by proclamation alone.[3] Of course, the tax was illegal as well as oppressive. The King's Treasury was also relieved in another way, namely, by the widespread and generally unrestrained exactions of the soldiers.[4]

[1] Carte, vol. i, p. 106, sq.   [2] Rymer's *Foedera*, 1743, vol. viii, pt. 2, p. 205.
[3] C.S.P., Ire., 1625-33, p. 343.   [4] Carte, vol. i, p. 116; Strafforde's *Letters*, vol. ii, p. 17.

At the beginning of the century the finances of Ireland were greatly embarrassed by the long war that had been raging, and by the generally unsettled and impoverished state of the realm, and, as we have seen, money had to be frequently sent from England to aid the Government in its difficulties. In 1603 the customs of the whole kingdom did not amount to fifty pounds a year.¹ This was caused by misappropriation by the corporate seaport towns, which refused to hand over to the King the customs which they had collected on his behalf. The first great administrative reform of Lord Chichester was the abolition of this abuse, which he successfully achieved in a few years, despite the vigorous resistance of the corporations.²

In 1606 the state of the revenue was still a constant subject of complaint. "Considering the great waste of the provinces where composition is yielded to His Majesty, especially Connacht, where very near £3,000 per annum is lost for want of inhabitants, and likewise the exceeding scarcity of money, it will fall out that at the most there will not be paid into His Majesty's receipts yearly above £14,000 or £15,000 English."³

Many suggestions were made during the next few years for the improvement of the revenue,⁴ and as a result of the continual exertions of Davies and Chichester an improvement began to be noticed about 1610. "His Majesty's revenues," wrote Davies to Salisbury in that year, "both certain and casual, are answered in due course; His Majesty's compositions in Leinster, Connaught and Munster are all revived and re-established; the customs in all the port towns which for many years have been first subtracted are now reduced."⁵ "I forbear to speak of the increase of the revenue of the Crown both certain and casual," wrote Davies three years later, "which is raised to a double proportion at least over what it was by deriving the public justice into all parts of the

---
¹ Carte, vol. i, p. 98. ² C.S.P., Ire., 1606-8, p. 213; C.S.P., Ire., 1611-14, p. 194.
³ C.S.P., Ire., 1606-8, p. 35. ⁴ C.S.P., Ire., 1611-14, p. 314. ⁵ C.S.P., Ire., 1608-10, p. 452.

realm; by settling all the possessions both of the Irish and English; by re-establishing the compositions; by restoring and resuming the custom; by reviving the tenures in capite and knights service; and reducing many other things into charge."[1] In the decade 1604-14 the revenue doubled.[2]

But, in spite of this improvement, the revenue was still not equal to the expenditure, and recourse was had to Parliament in 1613, when a subsidy was granted at the rate of two shillings and eightpence in the pound payable on the value of all moveable property, and four shillings on the value of all lands. The very poor were exempted from this tax, as it was provided that no part of the subsidy should be levied on any person possessing goods less than the value of three pounds, or lands of less than the yearly value of one pound.[3] This subsidy was somewhat of an innovation in Ireland, although it was modelled on an English tax which had been long in operation. "Subsidies had often been granted by Parliament in Ireland, ... yet these subsidies were not granted out of lands and goods, as the use of this time was in England, but a certain sum was imposed upon every ploughland after the ancient manner of taxes in England called hydage and carucage."[4] The incidence of the subsidies was sometimes a matter of complaint. "When any subsidy is there granted by Parliament the native and prime gentlemen be taxed in less sums than some of the meanest yeomen and freeholders, and this proceeds from the partiality of the commissioners appointed for the cessing of the subjects, who be always of the nobility and principal gentry of the shire, and make up a round subsidy, for which in their county they rate poor husbandmen higher in the subsidy books than some freeholders of good estates."[5]

During the remainder of the reign of James I. the revenue continued to improve; at the end of his reign

[1] *Discoverie*, p. 712. [2] C.S.P., Ire., 1611-4, p. 480.
[3] 10, 11 & 12 Jas. I, c. 10. [4] Carte, vol. i., p. 125.
[5] *Advertisements for Ireland*, MS., T.C.D.

the customs were farmed for close on ten thousand pounds a year;[1] and an important new source of revenue was created by the setting up of the Court of Wards and Liveries.[2] Attempts were also made to retrench expenditure by reduction in the army;[3] and many financial reforms were effected by Charles I., who was a better economist than his predecessor.[4]

The following detailed account gives a good idea of the state of the revenue in 1628[5]:—

ESTIMATE OF REVENUE (1) CERTAIN.

	£	s.	d.
Rents	20,470	1	6
Composition	8,229	13	4
Licence for transportation of	207	6	8
(2) CASUAL.	28,907	1	6
Court of Wards	7,000	0	0
Respite of Homage	200	0	0
Customs	11,451	0	0
Import of Wines	1,400	0	0
Seal and Writs	377	0	0
Felons' Goods	200	0	0
Green Wax Money	4,000	0	0
Star Chamber Fines	200	0	0
Spirits and Twentieth Parts	1,500	0	0
Fines upon Latitats	80	0	0
Clerk of the Faculties	20	0	0
	£26,480	0	0

However, the expenditure for that year was still greater than the revenue, and the deficit was made up by means of the Graces, and the illegal proclamation to which we have already referred. These additional revenues carried on the Government till 1634, when at last it was once again driven to asking Parliament for assistance.

The response to this request was the grant of several liberal subsidies. First of all, two subsidies at the same rate as those granted to James I. were granted,[6] and these were followed by four more subsidies at the same rate.[7] In the same year eight subsidies payable by the prelates and clergy were granted.[8] The six subsidies payable by the temporalty were " reduced to a certainty " by

---
[1] Carte, vol. i., p. 98. [2] Carte, vol. i, pp. 99-100; Leland, vol. ii., p. 476.
[3] Carte, vol. i., p. 100. [4] Carte, vol. i, p. 102. [5] C.S.P., Ire., 1625-33, p. 418.
[6] 10 Car. I, c. 1. [7] 10 Car. I, c. 2. [8] 10 Car. I, c. 23.

the House of Commons; the amount to be raised by the first four was £4,000, and by the last two £40,000; and the subsidies were apportioned in the counties as follows :[1]

	£		£
City Dublin	1,000	Fermanagh	900
County Dublin	1,000	Cavan	1,100
Kildare	1,150	Donegal	1,100
Louth	605	Londonderry	1,000
Queen's	800	Tyrone	1,200
Carlow	525	Antrim	1,200
Wexford	1,284	Monaghan	1,100
Wicklow	580	Down	1,300
County Kilkenny	1,400	Armagh	1,080
City Kilkenny	245	Town Carrickfergus	20
King's	900		
Longford	650		10,000
Meath	1,660		
Westmeath	1,050		
Town Drogheda	155		
	13,000		

	£ s. d.		£
Clare	1166 13 4	County Limerick	1,349
County Galway	1,366 13 4	City Limerick	524
City Galway	433 6 8	County Tipperary	3,152
Mayo	1,166 13 4	County Kerry	874
Sligo	900 0 0	County Waterford	756
Leitrim	700 0 0	County Waterford	606
Roscommon	1,066 13 11	City Cork	750
		County Cork	3,189
	6,800 0 7		11,200

It was one of Strafforde's dearest ambitions to put the Irish revenue on such a footing that Ireland would be able to pay the expenses of its own government, and produce a surplus for transmission to England. He consequently devoted much attention to the reform of the Irish finances, and before he left the country the customs were farmed for over £15,000 a year.[2] He succeeded in balancing Irish revenue and expenditure so that in 1641 the country was financially self-supporting,[3] the annual revenue being about £85,000, and the expenditure about £80,000.[4]

[1] *I.C.J.*, vol. i., pp. 104-7.   [2] Carte, vol. i. p. 166 ; Strafforde's *Letters*, vol. ii, p. 8.
[3] C.S.P., Ire., 1633-47, p. 299.   [4] C.S.P., Ire., 1633-47, p. 267.

The following is a detailed account of the revenue in 1640[1]:—

Certain Rents	£22,000
Composition	6,283
Court of Wards	10,000
Customs and Import	30,000
Tobacco Monopoly	5,000
Respite of Homage	300
Profits of the Seal and Original Rents	350
Felons' Goods	150
Green War Money	4,000
Faculties	50
Fines in Castle Chamber	800
First Fruits and Twentieth Parts	1,000
Fines upon Latitats	70
Mines	600
Licences for Ale	1,500
Licences for Yarn	1,000
Licences for Wine and Aquavitæ	1,200
	84,303

In 1640 Parliament granted four more subsidies at the same rate as before.[2] The first of these subsidies realized the sum of £46,170, although it was collected with the utmost laxity;[3] but the Houses of Commons passed resolutions which made a serious difference in the mode of assessment of the others, with the result that the second and third subsidies only amounted to £23,768,[4] and the fourth was never collected owing to the outbreak of the Rebellion.

During the period under review the revenue was farmed out, and it is certain that a great many abuses and oppressions took place in the collection.[5] Moreover, the revenue, when collected, was completely in the control of the Government, as there was no system whatsoever of parliamentary control at that date. The principal items of expenditure were the army and the civil establishment, but even at that early date there were many complaints of the pensions which were charged on the Irish revenue.[6] No part of the revenue was spent either on the improvement of the country or the encouragement of industry. We read that an almoner was appointed in 1616 to

[1] C.S.P., Ire., 1647-60, p. 235. [2] 15 Car. I, c. 13. [3] Carte, vol. i, pp. 205-213.
[4] Carte, vol. i, pp. 205-214. [5] *Advertisements for Ireland*, MS., T.C.D.
[6] C.S.P., Ire., 1611-14, pp. 197, 411; Leland, vol. ii, p. 476.

distribute the money assigned to charitable purposes;[1] but one cannot help suspecting that the appointment was of more benefit to the distibutor than to the recipients of the Royal alms. The execution of public works was the subject of much peculation and dishonesty. "Collectors of money there assessed on the country for performing of public works, as be, repairing of churches, gaols, bridges, town walls, and the like, should be called to account, for the country complains they are much burdened that way, and the works are not repaired, but left unfinished."[2]

(15) *Coinage and Credit.*

The coinage of Ireland was in a terrible condition of debasement at the beginning of the seventeenth century, owing to the monetary policy of Elizabeth, who systematically reduced the value of all Irish money with the object of ruining the commerce of the native Irish, and thus impoverishing and weakening the forces hostile to the Crown. The Elizabethan policy is well and clearly described by Carte[3]:—

"This cost Queen Elizabeth about one million two hundred thousand pounds, and the expense would have been much greater, and the success less, were it not for a new and extraordinary stroke of policy which she used on this occasion, and which contributed as much to that effect, as the strength of her army, the regularity of its pay, the continual supply of provisions, and the continuance thereof in its full strength for so many years together. This great Queen, who was the mother of our commerce, and who made it the business of her life to encourage and extend it as much and as far as was possible, had for the advancement thereof, in the beginning of her reign, been careful to put the coin of England on a good foot, reforming the standard of it, which had been much abased in the time of her father, Henry VIII. This was scarce more to the advantage of her subjects in England,

[1] *Buccleugh MSS.*, vol. i, p. 177. [2] *Advertisements for Ireland*, MS., T.C.D.
[3] Carte, vol. i, p. 23.

than the contrary method was serviceable for reducing the rebels in Ireland; where they used the same coin, only it passed then, as it does now, at somewhat an higher value there than it did in England.

"She had observed that the regular payment of her army in Ireland in gold and silver coin, drained a vast deal of the specie in England; and it being carried into that country in such quantities, that a great part of it, by inroads and plunder, or else by traffic for provisions and other ways of commerce, fell into the hands of Tyrone and the rebels, they were thereby enabled to procure from France, Flanders, Holland, and other countries, whatever arms, ammunition and provisions they wanted; and the King of Spain was also encouraged to send them a body of auxiliaries. To prevent this inconvenience, the Queen caused a base sort of money to be coined in the Tower of London, and by proclamation, 23 May, 1601, ordered it to be current in Ireland, and to be taken there in all payments as sterling money. Great quantities of this mixed money were sent over, and presently after no other coin was to be seen there."

In order to give full effect to this policy, all other money current in Ireland at the time was annulled and esteemed simply as bullion, and the export of English money to Ireland was prohibited.[1] An exchange office was set up and orders were given that all good coin should be brought in and exchanged for bad.

In addition to the export of good coin which this policy was intended to effect, other unpleasant results also attended it. The price of all commodities immediately increased fourfold, to the great detriment of all who were in receipt of a fixed income either from wages or rents of long leases.[2] When leases of lands fell in, the landlord would naturally refuse to renew at the old rent, and rents on new leases were then quadrupled to the great distress of the tenants in after years when the coinage

[1] Simon, *Irish Coins*, p. 28. [2] Simon, pp. 39-40.

was restored to its original value.[1] Indeed, the result of the debasement was to impoverish all classes of the community except the persons to whom patents were granted for exchanging the old coin for the new.

One of the first acts of James I. after his accession was to put the Irish currency on a new and satisfactory basis. A proclamation was issued in 1603 ordering the making of new shillings containing nine ounces of silver, and other coins of proportionate value, and making provision for a new coinage of copper money, which, however, should not be legal tender for more than the fortieth part of any payment. At the same time it was declared that the debased Elizabethan shillings should only pass for fourpence.[2]

It was difficult in practice to remove the ill effects of the debased coinage at one stroke. " This state," wrote Davies to Cecil in 1603, " doth suffer that punishment that Tantalus suffers in Hell; that is, it pines on plenty, for though there be no want of corn, no want of cattle, no want of anything necessary for the life of man, yet because of the want of *mensura publica*, which is money, to measure the price of all these things, they that want these things cannot have them; for the money which is now current by proclamation, albeit the same has decried from 12d. to 4d., yet it carries such with the people that they will not have it for the basest commodity at any rate. People will not give more than 2d. for the shilling, albeit there be 3d. worth of fine silver in it. And therefore, in the opinion of some, it were more profitable and more honourable for the King to resume the money at the same rate; for which he shall have more prayer and praises and acclamations in this poor Kingdom than for the 20,000 pardons he hath given."[3]

This general lack of confidence had also the effect of still further depleting the stock of coin in Ireland, as merchants bought up large quantities of the debased

---

[1] Simon, p. 40.   [2] Simon, p. 43; C.S.P., Ire., 1603-6, pp. 499-500.
[3] C.S.P., Ire., 1603-6, p. 112.

shillings at twopence each, and exported them abroad.¹ In order to put a stop to these evils, the value of the Elizabethan shilling was reduced still further to threepence in 1605.²

During the remainder of the period under consideration three distinct currencies circulated in Ireland. First, there were the Irish coins, properly speaking, which were made up of the coinage of James I., and what remained of the debased issue of Elizabeth. In the second place, English money was current at a third higher value than it bore in England, the English sovereign being worth twenty-six and eightpence in Ireland.³ The greater part of the copper and brass coinage was common to England and Ireland, and circulated in both countries at the same value.⁴ At one time Ireland was flooded with these copper coins, but a proclamation of 1635 provided that they should be no longer legal tender.⁵ Lastly, foreign coins circulated in Ireland with value according to the quantity of precious metal they contained. "The Spanish gold and silver is the coin that most aboundeth, and is chiefly reckoned on in that nation, especially Connacht and Munster."⁶ For the purpose of weighing foreign coin, special money weights were provided, and the manufacture of these money weights was made a monopoly.⁷ The system which prevailed at a later date of fixing the value of foreign coin by proclamation had not generally come into practice, although we find one such proclamation in 1641.⁸

Throughout the whole of the first half of the seventeenth century there were complaints of the scarcity of coin in Ireland. One result of the shortage was that money could not be borrowed except at an exorbitant rate of interest. "People of all sorts are driven to great extremity: for the better sort having occasion to take up money to serve their use are found to give to the greedy usurer £40 per £100, and that upon a pawn either

---

¹ C.S.P., Ire., 1603-6, p. 123.   ² Simon, p. 23.   ³ Simon, p. 44.   ⁴ Simon, p. 44.
⁵ Simon, p. 45.   ⁶ C.S.P., Ire., 1600-1, p. 126.   ⁷ Rymer's *Foedera*, vol. xix, p. 389.
⁸ C.S.P., Ire., 1633-7, p. 315.

of plate or land in mortgage, not daring to trust one another upon their bonds; the poorest being forced to pawn their apparel or other necessary implements wherewith they get their living and pay ordinarily for 20s., 6d. every week to their undoing. The poor farmer, is forced to sell part of his corn on the ground before it is ripe, for want of money to get in the rest of his corn. The want of money in the Kingdom is such that the poor want relief, as not being able to bring unto themselves any relief at all. All the coin that comes forth of England is of so pure silver that it is worth the value it goes for in any place whatsoever, and is therefore transported."[1] "The Dutchmen export store of coin from thence for their commodities. ..... The Irish merchants that frequent Spain bring thither every year vast quantities of Spanish coin to the East India Company of whom they receive a certainty in the pound for the Exchange. ... The Flemings and other strangers carry from Ireland the money that is had for their commodities, and this breeds the scarcity of money there as they pay ordinarily to the broker or usurer, thirty in the hundred interest and mortgages of lands be it the same rate or more; these brokers and usurers are the very mouths of the Commonwealth."[2]

The scarcity of coin was rendered worse about 1620 by the falling off in the amount of tillage, as great quantities of coin used to be brought into the country in exchange for the corn exported.[3] The practice of engrossing the coin, which is inevitable in any country where the money is not sufficient to meet the ordinary requirements of traders, went on as late as 1641.[4] "Nothing," complained Wentworth in 1637, "so much as want of money hath hitherto or doth at present hinder the growth of this Kingdom in the value of their lands, in their traffic, and in their manufactures."[5]

[1] C.S.P., Ire., 1608-10, p. 243.   [2] *Advertisements for Ireland*, MS., T.C.D.
[3] C.S.P., Ire., 1615-25, p. 425; C.S.P., Ire., 1625-32, p. 48.
[4] C.S.P., Ire., 1633-47, p. 261.   [5] Strafforde's *Letters*, vol. ii, p. 86.

The erection of a mint in Ireland was the subject of constant discussion. In 1623 we read of a "mint now set up or presently intended in Dublin,"[1] and in 1634 Parliament took into consideration the evil of the transportation of foreign gold and silver out of the Kingdom, and proposed the establishment of a mint. The King assented to this proposal, but did nothing in fulfilment of his promise, as we again find Parliament asking for a mint in 1641.[2] In 1637 a proclamation abolished the "title or name of Irish money or harp," and provided that all accounts and payments should be reduced into sterlings and made in English money; but this was only meant to affect accounts, and made no change in the value of the actual coins.[3]

One result of the shortage of coin was to render the borrowing of money extremely difficult and expensive. The rate of interest was regulated by an Act of Parliament in 1634, which provided that thenceforth the maximum rate should be ten per cent.; and that brokers negotiating loans should be entitled to receive a commission of five shillings per cent. on the making of the loan, and twelvepence per cent. on each renewal.[4]

---

[1] C.S.P., Ire., 1615-25, p. 426.
[2] Simon, p. 75; Strafforde's *Letters*, vol. i, p. 386; vol. ii, pp. 133-4; Carte, vol. i, p. 286.
[3] Simon, p. 46. [4] 10 Car. I, ch. 22.

CHAPTER II.

THE PERIOD OF DESTRUCTION, 1641-1660.

ON the whole, the first forty years of the seventeenth century were marked in Ireland by a general progress towards prosperity, and it is probable that that progress would have continued had it not been interrupted by one of the most disastrous periods of bloodshed and anarchy in Irish history. "The Kingdom," said Carte, "had enjoyed a continued peace of nearly forty years, during which the ancient animosities between the Irish and the English seemed to have been buried, and both nations cemented and (as it were) consolidated together by neighbourhood and conversation, by intermarriages, alliances, consanguinity, gossipings, and fosterings, and by a continuous intercourse of acts of hospitality, service and friendship; lands had been improved; traffic increased; and the kingdom in general raised to a more flourishing condition than it had ever known."[1]

The terrible twenty years which followed 1640 saw the complete ruin of the fabric which it had taken so many years to construct. For eleven years without intermission Ireland was torn asunder by a savage and destructive war; and the nine years following the conclusion of the war were devoted to the allocation of the lands of Ireland to Cromwell's soldiers and adventurers, and to the banishment of the old Irish beyond the Shannon or beyond the sea.

[1] Carte, vol. i, p. 309; Hely Hutchinson, *Commercial Restraints*, p. 10.

THE ECONOMIC HISTORY OF IRELAND 01

From an economic point the most serious loss which Ireland suffered during these terrible years was the drain on her population. Neglected fields may be reclaimed and retilled; devastated towns may be rebuilt; and the decayed industries of a country may be revived; but it is the work of many years to replace a community of industrious peasants and skilful artisans who have perished in battle, or have been driven to foreign lands. Many years after the termination of the Rebellion, its dire effects on the population were still felt. " The want of trade in Ireland," wrote Sir William Temple, " proceeds from the want of people; and this is not grown from any ill qualities of the climate or air, but chiefly from the revolution of so many wars and rebellions, so great slaughter and calamities of mankind as have at several intervals of time succeeded the first conquest of this Kingdom in Henry the Second's time until the year 1653."[1] It is true that it is possible to imagine circumstances in which such a decimation might prove an economic advantage owing to the pressure of population on the means of subsistence, and the lack of facilities for expansion; but Ireland was not in such a condition at the beginning of the Great Rebellion. On the contrary, the country was, if anything, underpopulated. We have seen that, when it was proposed in 1641 to raise sixteen thousand men for the service of the King of Spain, the Irish House of Commons objected that there were " not enough men in this Kingdom to maintain agriculture and manufacture."[2] If this was the condition of the country in 1641, what must it have been in 1652, when, according to Petty, 616,000 people had perished by sword, plague and famine, and also, no doubt, by emigration? As early as 1643 subscriptions were started for assisting " poor pillaged persons " to go to England.[3] Nor did the drain on the population caused by the war cease with the war itself. In the seven years following

---

[1] *Essay on the Trade of Ireland*, 1673.   [2] *I.C.J.*, vol. i. p. 276.
[3] *I.C.J.*, vol. i, p. 314.

1652, forty thousand soldiers were exiled abroad,[1] and great numbers of the old Irish landowners were transported into Connacht. Moreover, a great number of children and women were sent abroad as slaves.

" While the Government was thus employed in clearing the ground for the adventurers and soldiers by making the nobility and gentry of Ireland withdraw to Connacht and the soldiery of Spain, ' where they could wish the whole nation,' they had agents actively employed through Ireland seizing women, orphans and the destitute to be transported to the Barbadoes and the Plantations in America. It was a measure beneficial, they said, to Ireland, which was thus relieved of a population which might trouble the planters; it was a benefit to the people removed who might then be made English and Christians; and a great benefit to the West Indies sugar planters, who desired the men and boys for their bondsmen, and the women and Irish girls in a country where they had only Maroon women and negresses to solace them. The thirteen years war from 1641 to 1654, followed by the departure of 40,000 Irish soldiers, with the chief nobility and gentry to Spain, had left behind a vast mass of widows and deserted wives with deserted families. There were plenty of other persons, too, who, as their ancient properties had been confiscated, had no visible means of livelihood. Just as the King of Spain sent over his agents to treat with the Government for the Irish swordsmen, the merchants of Bristol had agents treating with it for men, women, and girls to be sent to the sugar plantations in the West Indies. The Commissioners for Ireland gave their orders upon the governors of garrisons to deliver to them prisoners of war; upon the keepers of gaols for offenders in custody; upon masters of workhouses for the destitute in their care ' who were of an age to labour, or, if women, were marriageable and not past breeding '; and gave directions to all in authority to seize those who had no visible means of livelihood and

[1] Prendergast, p. 87.

deliver them to these agents of the British sugar merchants, in execution of which latter direction Ireland must have exhibited scenes in every part like the slave hunts in Africa."[1]

In 1658 transportation abroad was made the penalty for failure to transplant into Connacht.[2] In this way was Ireland drained of the best of its inhabitants—a loss which it would take many years to repair, if indeed it were reparable at all. Such was the depopulation of the country in 1656 that even the new planters were alarmed at the difficulty of procuring labourers to till their newly-acquired lands; it appeared as if the whole country was destined to be waste for many years, and much of it for many ages.[3] "Had it not been," wrote Sir William Temple in 1673, "for the numbers of the British which the necessity of the late war at first drew over, and of such who either as adventurers or soldiers seated themselves here on account of the satisfaction made to them in land, the country had been by the last war and plague left in a manner desolate."[4] It must be remembered also that loss of population was considered a much more grievous calamity in the seventeenth century than it is now, as a decreasing population was at that time looked on as the greatest economic evil that could befall a country. This did not make Ireland's position any worse than it would otherwise have been; but it enables us to realize the guilt of the statesmen who deliberately pursued such a policy.[5]

The period immediately following the conquest of Cromwell was characterized by a war of peculiar violence against Catholics. The Penal laws put into operation were a foretaste of the code of the eighteenth century. That is to say, they were laws not directed merely against the public exercise of the Catholic religion, or the elevation of Catholics to offices of public importance, but

[1] Prendergast, pp. 89-90. [2] Prendergast, p. 145.
[3] Lawrence, *The Interest of England in the Well Planting of Ireland with English*, Dublin, 1656. [4] Temple, *Essay on Trade of Ireland*.
[5] See the Introduction to *Hull's Edition of Petty's Works*, p. lxii.; Cunningham *Growth of English Industry and Commerce*, vol. ii, p. 387.

were laws designed rather to degrade the Catholic population, to rob them of ambition by closing to them many means of livelihood, and to deprive them of their property —laws designed " to make them poor and to keep them poor." Catholics were discriminated against in the Cromwellian settlement by being required to discharge a greater burden of proof to establish their " innocency " than Protestants,[1] and the net result of their failure to discharge the heavy burden thus imposed was that the majority of the Catholic landowners in Ireland were dispossessed. Catholics, moreover, were forbidden to reside or carry on any trade in a walled town.[2] Any Catholic who refused to take the iniquitous oath of abjuration prescribed by the Commonwealth Government was obliged to forfeit two-thirds of his possessions. As the majority of the Catholics in Ireland heroically refused to take this oath, which implied the abandonment of their religion, a great amount of Catholic property must have been confiscated.[3] It is, therefore, apparent that for a few years in the middle of the seventeenth century, the majority of the population of Ireland were subject to a degrading penal code, which, had it continued in operation for any length of time, would undoubtedly have had the most injurious effects on the economic position and character of the people.

Of course, during these years the ordinary agricultural operations of the country were, to a large extent, suspended. The greater part of the population were engaged in the war on one side or the other, and, even if not actually so engaged, were perpetually menaced by the soldiers of the different armies, and by the bandits whose existence is inevitable in a country in a state of anarchy. Naas was in the centre of a rich and prosperous county; its situation within the boundaries of the Pale which had always enjoyed a reasonable degree of security, together with its proximity to the great market of Dublin, should have rendered it the centre of a cultivated and

[1] Butler, *Confiscation in Irish History*, p. 129.
[2] Moran, *Persecutions of Irish Catholics*, p. 49.
[3] Moran, *op. cit.*, pp. 364-381.

opulent countryside; and yet in 1643 there was no tilled land to be found near or about Naas.[1] In the following year wheat was sold in Dublin at 22s. per barrel;[2] and in 1647 at 34s. a barrel in Cork.[3] "The quantity of corn that was brought into Ireland in one year after the last war is incredible, as they relate it. A person of good quality and fortune told me he was then a merchant, and that he sent great quantities of wheat and malt himself, that above 10,000 of our quarters in one year went to Waterford, and that to all parts there were no less than 100,000 quarters in one year."[4]

Agricultural capital and produce were destroyed wholesale. "Whosoever will undertake for the rescue or recovery of this Kingdom must necessarily furnish and provide a good stock of horses; for the horses of the Kingdom are destroyed between the enemy and us."[5] Each army systematically destroyed the cattle of the other; a practice based on sound military notions, but economically hard to recommend. "Even to the beasts of the fields and improvements of their lands is their malice towards the English expressed; for they destroy all cattle of English breed, and declare openly that their reason is because they are English."[6] So great was the destruction of cattle that in 1652, Ireland, the great cattle-exporting country, had to import cattle from Wales.[7] Petty reckoned that in 1641 the cattle and stock of Ireland was worth above four millions, but that in 1652 it was not worth more than £500,000.[8]

In 1642 the Lords Justices reported that they had had to "burn much corn to keep it from the rebels";[9] and in the same year reported that they found "a great want of mills here, in regard the rebels burnt and destroyed all the mills whereof we might have power, and our soldiers wasted the rest."[10] That the destruction of all agricultural wealth was a deliberately conceived part of the

---
[1] *Ormond MSS.*, n.s. vol. i, p. 60. [2] *Ormond MSS.*, n.s. vol. i, p. 82.
[3] *Egmont MSS.*, p. 421. [4] *The Character of the Protestants of Ireland*, London, 1689.
[5] *Ormond MSS.*, n.s. vol. i, p. 251. [6] *Ormond MSS.*, n.s. vol. ii, p. 35.
[7] Petty, *Political Anatomy*, p. 21. [8] *Ibidem.* [9] *Ormond MSS.*, n.s. vol. ii, p. 133.
[10] *Ormond MSS.*, n.s. vol. ii, p. 217.

Cromwellian plan of campaign is strikingly illustrated by a list of "military weapons" issued from the store at Waterford, wherein the following "military weapons" are included:—" Eighteen dozen of scythes with handles and rings, forty reaping hooks, and whetstones and rubstones proportioned."¹ The Commissioners of Ireland, writing to the Parliament of England in 1651, stated that Colonel Hewson had started with his troop for Wicklow, where he "doth now intend to make use of scythes and sickles that were sent over in 1649, with which they intend to cut down the corn growing in these parts."² For years corn was a scarce commodity in Ireland.³

The soldiers deliberately destroyed as much of the woods as they could;⁴ and most of the iron-works throughout the country were broken down.⁵ The valuable silver and lead mines at Dunally in County Tipperary, the success of which had been a project dear to Charles I., were utterly and completely laid waste; the accumulated stock of lead, coal, timber and tools worth £10,000 was taken away, and many of the workmen were killed.⁶ Buildings were razed to the ground on all sides; "the houses of Ireland in the year 1641 were worth 2½ millions; but in the year 1652 not worth one-fifth of the same."⁷ In the year 1641 a great many bogs were being reclaimed by a new and successful method, but this pre-eminently useful work was abandoned when the Rebellion broke out, and much partially reclaimed land was suffered to relapse into its original valueless condition.⁸

This combination of neglect and deliberate destruction produced a desolated country and a population on the verge of famine. As early as 1643 the Lords Justices wrote to England that if they were not speedily relieved they were "like to perish, for famine doth much threaten us."⁹ In the answer to one of the queries on the Cessation with Inchiquin in 1648 we read that:—" It is hopeless to

¹ Prendergast, p. 78. ² Moran, *Persecutions of Irish Catholics*, p. 241.
³ Dunlop, pp. 131-137. ⁴ Gilbert, *Confederation and War*, vol. iii, p. 112.
⁵ Boate, p. 117. ⁶ C.S.P., Ire., 1660-2, p. 153.
⁷ Petty, *Political Anatomy*, p. 21; *Ormond MSS.*, n.s. vol. ii, p. 20.
⁸Boate, p. 77. ⁹ *I.C.J.*, vol. i, p. 313.

get money from a country so totally exhausted and so lamentably ruined . . . . no common granaries for the public, and but very small store of grain with any private persons, in so great a dearth of corn as Ireland hath not seen in our memory,] and so cruel a famine which hath already killed thousands of the poorer sort."[1] The great want of corn was the subject of constant complaint.[2] In 1650, Ludlow spoke of the " poor wasted country of Ireland," adding that the Irish had always wasted the land by bad cultivation, and that they had of late been worse than ever " being in daily apprehension of being removed."[3] Some years late, Petty found the people living on potatoes, and the cultivation of that root must have been stimulated by the confusion of the years of the war. It was then and for many years later the practice to dig out the tubers just as they were wanted, as a crop so treated could not easily be carried away or destroyed.[4] In 1652, proclamations were issued prohibiting the slaying of sheep and lambs, wherein it was recited that starvation and plague were rife throughout the land.[5] In 1653 many people perished through sheer want in Wexford, formerly one of the most prosperous of Irish towns.[6] " The handful of natives left," we read in 1655, " are poor labourers, useful simple creatures whose design is only to support themselves and their families, the manner of which is so low that it is a design rather to be pitied than by anybody feared, envied, or hindered."[7] In another pamphlet of the same date the condition of the Irish is described as follows:—" They were strong; they are weak; they were numerous; they are consumed by sword, pestilence and famine; they were hearty; they are out of courage; they were rich; they are poor and beggarly; they had soldiers; they are left naked; they had cities; they have but cottages in which to put their

[1] Gilbert, *Confederation and War*, vol. vi, p. 271.   [2] *Ibidem*, vol. vii, p. 227.
[3] Bagwell, vol. ii, p. 300; see *Egmont MSS.*, pp. 496, 536.
[4] *Ibidem*; in 1657 potatoes were the ordinary food of the Irish peasantry; Cole, *Adam in Eden, or The Paradise of Plants*, London, 1657.
[5] Bagwell, vol. ii, pp. 245, 301.   [6] Dunlop, p. 329.
[7] Gookin, *The Great Case of Transplantation Discussed*, London, 1655.

heads.'"¹ In the same year we read of the "wasted desolate country."² In a few years three-fourths of the cattle in Ireland had been destroyed; tillage had utterly ceased; and the national revenue had sunk to the lowest sum it had reached for many years.³ For many years afterwards the country was infested with wolves.⁴

The following is Prendergast's terrible description of the condition of Ireland in 1652-3 :—" Ireland, in the language of Scripture, now lay void as a wilderness. Five-sixths of her people had perished. Women and children were found daily perishing in ditches, starved. The bodies of many wandering orphans whose fathers had embarked for Spain and whose mothers had died of famine were preyed upon by wolves. In the years 1652 and 1653 the plague and famine had swept away whole countries, that a man might travel twenty or thirty miles and not see a living creature. Man, beast, and bird were all dead, or had quit those desolate places. The troopers would tell stories of the places where they saw a smoke, it was so rare to see either smoke by day or fire or candle by night . . . . if two or three cabins were met with, there were found there none but aged men with women and children; and they, in the words of the prophet, ' became as a bottle in the smoke; their skins black like an oven because of the terrible famine.' They were seen to pluck stinking carrion out of a ditch black and rotten; and were said to have even taken corpses out of the grave to eat."⁵ " The whole country was wasted and consumed," we read, " so that for sometimes ten, sometimes twenty miles together, all the Kingdom over, a traveller should not behold either man, bird or beast, the very fowls of the air and wild beasts of the field being either dead or having forsaken those unfortunate desolations. Thousands of the Irish lay daily starving for want of food . . . . nay, the famine at last grew to that height that they not only fed upon horses, but upon dead bodies

---

[1] Gookin, *The Author and Case of Transplanting the Irish into Connaught Vindicated*, London, 1655. [2] Dunlop, p. 477. [3] Prendergast, p. 79.
[4] *I.C.J.*, vol. i, p. 539; O'Flahertie, *Chorographical Description of Iar Connaught*, 1684, Ir. Arch. Soc., vol. ix, p. 184. [5] Prendergast, pp. 307-8.

digged up again out of their graves.'"¹ "Each one's thoughts were solely devoted to preserve life and to avoid the impending destruction. Hence resulted a dearth of all articles of food, and with a famine, a pestilence, too, assailed us. Thus the three scourges of God, of which David had to choose but one—famine, war and pestilence—were at the same time inflicted upon us."²

In short, whatever may have been the position of the inhabitants of Ireland in 1641, they were thirteen years later "as miserable a people as their worst enemies could wish them,"³ having lost in the interval five-sixths of their possessions.⁴ The poverty of the old inhabitants may be judged from the *Book of Transplanters' Certificates*, which contains an inventory of the goods belonging to many of the landowners who were forced to transplant into Connacht. For instance, Lord Dunboyne owned four cows, ten garrons, and two swine; Dame Katharine Morris, one and a half acres of summer corn, sixteen garrons, ten cows, nineteen goats and two swine; Lady Mary Hamilton, three and a half acres of summer corn, forty cows, thirty garrons, forty-six sheep, and two goats; and Viscount Ikerrin, sixteen acres of winter corn, four cows, five garrons, twenty-four sheep and two swine.⁵ The shocking number of beggars in the streets of Dublin attracted attention in Parliament in 1661.⁶

The results arrived at by the "political arithmetic" of the seventeenth century are not usually of any great value as statistics, but the following figures from Petty,⁷ are worth quoting as showing the enormous scale of the damage wrought by the Rebellion. "The effects of the Rebellion were these in pecuniary value, viz.:—

By loss of people	£10,335,000
By loss of their superlucration of soldiers ...	4,400,000
By the superlucration of the people lost ...	6,000,000
By impairing the worth of lands	11,000,000
,, ,, of the stock ...	3,500,000
,, ,, of the housing ...	2,000,000
	37,235,000"

¹ *The Sad and Lamentable Condition of the Protestants of Ireland*, London, 1689.
² Missio Soc. Jesu. ³ *Ormond MSS.*, n.s. vol. i. p. 247.
⁴ *The Great Case of Transplantation Discussed*, London, 1655.
⁵ Prendergast, p. 105. ⁶ *I.C.J.*, vol. i, p. 406. ⁷ *Political Anatomy*, p. 22.

Of course, the small but growing trade of Ireland was also ruined. In 1642 the Lords Justices reported that all manner of trade in or out of Dublin had been stopped,[1] and a year later we read more complaints on the same subject :—" Few or none dare now come hither with any commodities, and indeed the merchants here, having all their remaining stocks and estates wrested from them by the State, are not possibly able, however so willing they are, to help us, so as now the little trade driven here is likely to be destroyed, and we shall not only fail in getting any supplies that way from abroad, which hath hitherto been a great means of our preservation, but our poor and mean quantity of native commodities in the few ports we have, which cannot be manufactured here in these times, though they are not considerable, cannot gain us returns."[2] Many commodities which did succeed in reaching a seaport could get no further want of shipping;[3] the sea was infested with pirates who attacked and pillaged merchant vessels trading to Irish ports.[4] The Commons Journal for 1661 contains a reference to the " universal decay of our trade,"[5] and the same complaint is made in the reports of the Cromwellian Commissioners.[6] The total customs and excise in the period 1649-56 only amounted to £252,074.[7]

Although Ireland was looked on as part of the Commonwealth, free trade between England and Ireland was not permitted. In this respect Ireland was put in a worse position than Scotland or the Plantations. In 1652 free trade was opened between Ireland and Scotland and the Isle of Man, but not between Ireland and England;[8] and two years later the adventurers in England refused to proceed to Ireland except on the undertaking that a free trade between England and Ireland would be allowed, and Cromwell gave them a reassuring

---

[1] *Ormond MSS.*, n.s. vol. ii, p. 75. [2] *Ormond MSS.*, n.s. vol. ii, p. 240.
[3] *I.C.J.*, vol. i, p. 363. [4] Dunlop, pp. 65, 126-7. [5] *I.C.J.*, vol. i, p. 392.
[6] Dunlop, p. 112; and see *Contributions Towards a History of Irish Commerce*, by Wm. Pinkerton, Ulster Archæological Journal, vol. iii, p. 189.
[7] Dunlop, p. 639. [8] Dunlop, p. 135.

answer to this request.¹ Nothing, however, was done, for, in the following year, a petition was presented to Henry Cromwell from the citizens and inhabitants of Dublin, wherein it was set out that Ireland was by Act of Parliament part of the Commonwealth; that numerous petitions had been granted for the abolition of the burden of customs payable on goods imported from England, which were often exacted two or three times; that in this respect Ireland was in a worse position than Scotland or any of the Plantations; and that repeated promises to reform the matter had been broken.² Again nothing was done, and in 1659 a precisely similar petition had to be presented.³ The one commodity which was allowed to be freely exported to other parts of the Commonwealth was linen,⁴ but with this exception, the general rule was that there was no free trade between England and Ireland, and that Ireland was thus treated worse than any other part of the Commonwealth.⁵ Until 1655 a general embargo prevented the exportation of all provisions to any part of the world.⁶

In the same way Ireland was discriminated against in the matter of her carrying trade. By the Navigation Act of 1654 it was provided that goods from the Plantations might be imported into England or Ireland only in ships belonging to England or the Plantation whence the goods came.⁷ Irish ships, however, were not allowed to carry goods from the Plantations even to Ireland, and, as a matter of fact, Irish shipping now disappeared from the seas. We have seen that an increase in the number of ships owned by Irish merchants had first been noticed about 1630,⁸ and the volume of Irish shipping was stated to have increased a hundredfold in the years 1630-40.⁹ Yet in 1665 Ireland was said to have no shipping of its own.¹⁰ In the same year the Lord Deputy and Council were requested to impress a thousand men in Ireland for

---
¹ *Egmont MSS.*, p. 542.   ² C.S.P., Ire., 1647-60, p. 572.
³ C.S.P., Ire., 1647-60, p. 678.   ⁴ C.S.P., Ire., 1647-60, pp. 572, 826.
⁵ Firth, *The Last Years of the Protectorate*, vol. ii. pp. 162-3.   ⁶ Dunlop, p. 539.
⁷ Firth & Rait, *Acts and Ordinances of the Interregnum*, vol. ii, p. 559.
⁸ *Egmont MSS.*, p. 72.   ⁹ C.S.P., Ire., 1633-47, p. 252.   ¹⁰ Carte, vol. iv, p. 236.

service in the English fleet, but were unable to do so—" albeit there are many maritime towns and ports in the Kingdom, yet so great is the desolation wrought here by the late rebellions, that those ports do for the most part lie waste, and the maritime towns are in a great measure depopulated of seamen. . . . The trade hitherto driven in this kingdom is not carried on by the shipping belonging to this Kingdom, but by shipping belonging some to His Majesty's subjects of England, and some to foreigners and strangers, Dutch and others."[1]

The coast fisheries were likewise ruined. "The fishery for pilchards and other fisheries formerly here are now discontinued, through the paucity and poverty of our people."[2] One factor which probably contributed to cause this was the clause in the Navigation Act of 1654 which provided that no fish should be imported into England except in English vessels.[3]

The towns which were the chief centres of whatever trade was carried on in the country fell into decay. In 1645 Dublin was partially in ruins, and a proclamation in that year forbade soldiers to pull down the deserted houses to make fires with the timber.[4] In 1647 Lord Inchiquin was charged with having given houses in the city of Cork and farms in the suburbs to his own menial servants. His answer was that, on expelling the Irish out of Cork, it was to the benefit of the State that he should place any persons in the houses on the sole condition of upholding them, which otherwise being waste and uninhabited would have fallen to the ground; and though by this means many of the houses were preserved, yet for want of inhabitants about ten thousand good houses in Cork and as many in Youghal had been destroyed by the soldiers.[5] All the old Irish inhabitants were driven from the towns; and

[1] C.S.P., Ire., 1662-6, p. 524.   [2] *Ibidem.*
[3] Firth & Rait, *Acts and Ordinances of the Interregnum*, p. 560.
[4] Prendergast, p. 274.
[5] Gething, *Articles Humbly Presented to the House of Commons against Murrough O'Brien, Lord Baron of Inchiquin, with a Clear Answer thereto made*, London, 1647.

whatever trade continued to be carried on was conducted by factors on behalf of English merchants.[1]

Ireland during this period ceased to be able to support its own Government as it had done during the later years of the reign of Charles I. During the period of chaos following the outbreak of the Rebellion, the whole subject of the public finance of Ireland is obscure. "From the distracted state of the Government which prevailed from 1641 till the settlement under Cromwell, no estimate can be formed of the various taxes or even an enumeration of the objects which composed them."[2] During the ten years 1651-1660 great sums of English money were remitted to Ireland to defray the expenses of Government, and the revenue of Ireland always fell far short of the expenditure. In 1653 the revenue was £197,000, and the expenditure £630,814; and in the following year the army was estimated to cost £47,000 a month, whereas the total revenue which could be raised in the country was not more than £10,000 a month.[3] The principal source of revenue was the monthly assessment, which was at first fixed at £10,000, but afterward raised to £12,000.[4] The following abstract of all moneys received and paid for the public service in Ireland from July, 1649, to November, 1656, shows that Ireland was far from being financially self-supporting[5]:—

RECEIPTS.

Transmitted from England		£1,566,848
RECEIVED OUT OF IRELAND.		
Assessments of Ireland	£1,309,695	
Rents of forfeited and sequestered lands, houses, etc.	161,598	
Rents of impropriate and sequestered tithes	135,524	
Customs and Excise	252,474	
Casual Revenue	83,254	
Total Revenue ...		1,942,545
		3,509,393
Total Expenditure		3,485,170

[1] Prendergast, p. 278. The clearance of the towns is fully dealt with in Prendergast, p. 275, et. seq.; and see *Kilkenny Journal of Arch.*, n.s. vol. iii, p. 326.
[2] Clarendon, *Revenue of Ireland*, 1791. p. 5.
[3] Firth, *The Last Years of the Protectorate*, vol. ii, pp. 164-5.
[4] *Ibid.*, p. 165.   [5] Dunlop, pp. 639-45.

In the first year after the Restoration the total revenue of Ireland from all sources was only £42,000.[1]

Another result of the Rebellion was the debasement of the coinage, which had been put on a fairly satisfactory basis in the earlier years of the century. In 1642 great quantities of plate and bullion were brought in to be coined,[2] and the erection of a royal mint in Dublin was again advocated.[3] Coins of silver and copper were also issued by the Confederates at Kilkenny.[4] During the next ten years the coin of the country was for the most part carried abroad, and Ireland in 1651 suffered from an extreme shortage of currency. In 1652 complaint was made of the great need of good English money which was felt; nothing circulated but debased and worthless Perus,[5] which had been brought to Ireland, together with quantities of clipped English money, by London merchants, some of whom were afterwards executed for this fraud.[6] So great was the quantity of clipped money in circulation that a proclamation was issued that all money, even English, should be given only by weight.[7] Two years later the same complaint was being made; any remnant of good English money that remained was being exported, and its place was being taken by base Perus.[8] In the following year we read of the " spreading of base Perus and other like money, which, like a cancer, hath eaten out all the good."[9] "Like a gangrene this adulterate coin spreads far and near; it banishes hence the current coin of Spain and eats up the good English money which the merchants make it a secret trade to export into England or into foreign parts. .... Little other money is visible save this counterfeit American, which ordinarily goes for 4s. 2d, and upon assay is found not to value 2s. 4d., and most of that which is current is little better than brass or alchemy."[10] The country, moreover, was plagued with counterfeit money.[11]

[1] Carte, vol. iv, p. 100. [2] Simon, p. 46. [3] Simon, p. 47.
[4] *Kilkenny Archæological Journal*, vol. i, p. 442; vol. iii, pp. 16, 67.
[5] Dunlop, p. 248. [6] Simon, p. 49. [7] Simon, p. 49. [8] Dunlop, p. 402.
[9] Dunlop, p. 557. [10] Dunlop, pp. 592-3. [11] Dunlop, pp. 402, 551, 657.

The erection of a mint in Ireland was frequently discussed, and arrangements were made for its establishment; this, however, was put off indefinitely by reason of "multiplicity of business."[1] The need of small money for every-day transactions was met by the practice of tradesmen issuing tokens, which circulated locally. Great numbers of these tokens were issued under the Commonwealth.[2]

To recapitulate: the twenty years from 1641 to 1660 were years of wholesale loss and destruction. Every department of economic life suffered equally; the land which had so greatly improved in the preceding forty years, was laid waste; tillage all but ceased; and the cattle of the country was insufficient to supply even the Irish demand. The woods were felled; the promising mines were neglected and let fall into decay; and the tracts of bog and mountain, which were being reclaimed with so much labour, were suffered to fall back into their original condition of unfertility. Trade and industry languished; the shipping of Ireland disappeared from the seas; and the wealth derivable from the fisheries was again abandoned to foreigners. The penal laws, moreover, and the commercial restraints, were a foretaste of what was to happen in the eighteenth century. But more important than anything we have enumerated was the dispossession and degradation of the Catholic landowners, and the plantation of the country anew. This feature of the Cromwellian policy was far-reaching in its consequences, and left its mark so deeply on the later history of the century that we deal with in in the next chapter, which treats of the period following the Restoration.

---

[1] Dunlop, pp. 301, 316, 324, 402.
[2] Proc. R.I A., vol. iv, p. 345, and Appendix iv.; *Kilkenny Archæological Journal* vol. ii, p. 155; Lindsay, *Coinage of Ireland*, Appendix 3.

CHAPTER III.

THE PERIOD OF RECONSTRUCTION, 1660-1689.

(1) *Generally.*

THE period which elapsed between the Restoration and the Revolution was, in some respects, very similar to the period of construction in the earlier part of the century. Again, Ireland started to build up a new economic life from ruins of that which had been overthrown in a period of warfare. As in the earlier case, the power of recuperation displayed was remarkable, and the process of reconstruction was rapid; but in a few years the whole edifice was again destroyed by another outbreak.

The first branch of economic life which we shall have to consider in this chapter is the resettlement of the ownership and tenure of land. The Cromwellian war was followed by confiscation on a scale unprecedented in Irish history, and the great majority of the old freeholders were either driven to emigrate, or degraded to a lower position in the social scale. The levelling down process, however, was accompanied by one of levelling up, and the same scheme of settlement that caused such distress amongst the landlords, conferred certain undoubted benefits on many of their tenants by elevating them to the position of freeholders. Those who remained tenants, and the new tenants who came into existence after the Restoration, still enjoyed a great degree of immunity from the evils which, at a later date, so grievously oppressed the Irish tenantry; the "custom of the country" was generally recognised, as in the earlier part of the century; and a

THE ECONOMIC HISTORY OF IRELAND 117

tenancy from year to year was still felt to confer a high degree of security. Lowest in the scale of those who lived upon or by the land, the cottiers remained the fixed substratum of the population, affected less than any other class, either for better or worse, by the Cromwellian settlement.

In the sphere of trade the period now about to be considered was pregnant with change. We have seen that, on the whole, Irish trade was free and unrestrained during the earlier half of the seventeenth century, but it was now destined to experience the first effects of the later policy of trade restriction. Soon after the Restoration, England embarked on her colonial policy, which is best known by the provisions of the Navigation Acts, whereby she sought to secure, principally for naval reasons, the whole benefit of the Plantation trade, to the exclusion of foreign countries, and of Scotland and Ireland. As a matter of fact, Ireland did not suffer any appreciable damage as the immediate result of these acts; her Plantation trade was small at the time. The effects of the Navigation Acts in the eighteenth century were extremely injurious to Irish prosperity, and from that point of view they were amongst the most important events of the latter half of the seventeenth century; but their ill-effects did not show themselves seriously in practice during the period now under consideration.

More important than the Navigation Acts were the measures adopted by England to keep Irish produce out of the English market. Immediately after the Restoration high import duties were placed on all manufactured goods entering England from abroad, no exception being made in favour of Ireland, and these had the effect of depriving Ireland of all hope of obtaining a sale for her manufactured products in England. However, in practice, this was not a very serious deprivation; the industries of Ireland at the time were small and undeveloped; and the amount of manufactured goods prepared for export was negligible. The principal article of export to England, on the other hand, was live cattle. If the trade in this commodity had been allowed to continue undisturbed, a considerable income would have been derived from it by the proprietors of Irish lands, and at the time that it was suppressed, it constituted three-fourths of the total external trade of Ireland. However, the English Parliament, animated nominally by a fear of a fall of rents on English breeding lands, but really by jealousy of the Duke of Ormond, determined to put a stop to this promising trade, and by a series of Acts passed shortly after the Restoration prohibited the import into England of all Irish cattle, sheep, pigs, and provisions.

The passing of these measures is an illustration of what has been pointed our earlier in this book, that the real enemy of Irish prosperity was not the English king, but the English Parliament. Charles II. was fully convinced of the unwisdom of the Cattle Acts, both from the English and Irish points of view, and did all in his power to prevent them from passing into law; but Parliament was quite determined to pass them, and was in a position to do so in spite of the king, who was in need of a Parliamentary grant of money at the time.[1] These Acts caused much immediate distress in Ireland, but ultimately they proved a benefit in disguise. Instead of exporting their cattle alive to England, the Irish began to kill them at home, and to make them up into provisions, which they exported abroad. In this way was laid the foundation of the prosperous provision trade, which attained to such large proportions, and which became, in the eighteenth century, the staple trade of the greater part of the country. Another result of the Cattle Acts was that in many places sheep took the place of cattle on the pasture lands, and thus increased the stock of wool, giving thereby an impetus to the woollen industry. Large quantities of this wool, moreover, were smuggled to foreign countries, where it was mixed with foreign wool, and worked up into the cloths and serges that in a few years competed seriously

[1] Carte, vol. iv, pp. 242, 262, 287.

with, and severely damaged, the English woollen manufacture. Thus, Ireland was the ultimate gainer, and England the ultimate loser, by the Cattle Acts.

During the period treated in this chapter, Irish industries were encouraged and advanced, both by the private enterprise of the Duke of Ormond, and by the legislation of the Irish Parliament. The woollen industry grew considerably, so that a large quantity of woollen goods was beginning to be exported; and the linen industry also showed signs of development.

On the whole, then, the period from the Restoration to the Revolution was one of economic progress. Of course, things did not mend immediately after the restoration of settled government; for some years there were great complaints of poverty; food was dear; farmers were obliged to abandon their farms because of the shortage of labour; and great numbers of cattle died of the plague.[1] But soon the marvellous recuperative power of Irish society began to operate; from about 1670 the country rapidly improved, and from that date until the Revolution no exceptional distress was felt except in one or two years—such as 1674 and 1687—when bad harvests created a temporary depression.[2]

Land values rose as security increased. "I have heard," wrote Sir John Temple in 1685, "of more buying and selling of land within the last month than I did within a year before, and very little is now sold for less than fifteen years' purchase."[3] The population increased, and was not depleted by emigration; on the other hand, the country was enriched by the advent of many industrious foreigners. The Catholic population were not hampered by any degrading penal laws. Manufacturing industry flourished; the revenue increased yearly; and the Government was carried on without resort to any new or extraordinary taxation. "Ruined though Ireland was,"

[1] C.S.P., Ire., 1647-60, p. 719; C.S.P., Ire., 1662-6, pp. 530, 644, 656; C.S.P., Ire., 1666-9, pp. 109, 131, 257, 397, 709.
[2] C.S.P., Ire., 1666-9, p. 425; C.S.P., Dom., 1673-5, p. 323; *Dartmouth MSS.*, vol. iii, p. 119; *Ormond MSS.*, vol. vii, p. 493.   [3] *Ormond MSS.*, vol. vii, p. 409.

says Professor Cunningham, "at least it had rest; under Charles II. a semblance of civil order was maintained, and the country showed signs of beginning to enjoy a period of comparative prosperity."[1]

The prosperity and wealth of Ireland, then, increased considerably during the reigns of Charles II. and James II., and they would have continued to increase still more rapidly, if the policy of the latter king had been allowed to take effect. Whatever may be thought of the composition or behaviour of the famous Patriot Parliament, it must certainly be admitted that it is weighty evidence of the wisdom and public spirit of that body that, meeting as it did, at a troubled time, it immediately devoted its attention to the amelioration of Irish economic conditions. One cannot help regretting that this wise legislation was not destined to take effect. There is little doubt that, had not the Williamite Rebellion been a success, Ireland in the eighteenth century would have been a much more prosperous and much more contented country than it actually was.

One of the first acts of the Patriot Parliament was entitled, "An Act declaring that the Parliament of England cannot bind Ireland";[2] in other words, an Act asserting that legislative independence which was afterwards withheld for nearly a hundred years. If this Act had not been repealed, the worst measures of commercial repression in the eighteenth century would have been impossible. The English Acts prohibiting the export of Irish woollens and glass were effective only because the legislative independence of Ireland was not fully recognised; moreover, the Irish Parliament would have been free to adopt measures to protect and encourage Irish industries without the interference of the English Privy Council. If this Act had not been ignored, Ireland would have enjoyed, during the whole of the eighteenth

[1] *Growth of English Industry and Commerce*, vol. ii, p. 367.
[2] Davis, *Patriot Parliament*, p. 43.

century, the prosperity and progress, which, as it was, she did not experience until after 1782.

Another Act of the Patriot Parliament conspicuous for its wisdom was one dealing with taxation. It provided that a subsidy should be raised by means of a land tax, which should in the first instance be paid by the occupier, who should be allowed to deduct the tax from his rent when the land was let at the whole of its value; and that when the land was let at half its value or less, the tenant should pay half of the tax. Thus, not only rack-rented farms, but all farms let at any rent, no matter how little, over half their value, were free of this tax. When, Davis asks, has a Parliament of landlords in England or Ireland, in distracted or in quiet times, acted with equal liberality?

The bitter and harmful tithe war which did so much to hamper economic progress in after years would also have been avoided, had the Patriot Parliament's laws been maintained. The settlement of the tithe question arrived at by that body was simple and satisfactory, namely, that Protestants should pay their tithes to the Protestant clergy, and Catholics to the Catholic clergy.[1]

A statute was passed for the encouragement of the Irish coal mines, whereby the importation of English, Scotch, and Welsh coals into Ireland was forbidden. Nor was profiteering to be allowed: "And for as much as the owners or proprietors of the coal pits of Kilkenny and other coal mines in this Kingdom may upon passing of this Act, enhance and raise the price of coals to the defeating of the ends proposed hereby; be it therefore enacted that no owner or proprietor of such coal pits or coal mines or seller of coals at any of the said pits shall at any time hereafter receive or demand more than ninepence for each barrel of coals, Bristol measure."[2]

The most remarkable of all these Acts for the benefit of trade, was one entirely removing all restraints on

[1] Davis, *Patriot Parliament*, p. 51.
[2] Davis, *Patriot Parliament*, pp. 52-54.

Irish trade with the Plantations, and for the encouragement of Irish shipping. This Act swept away all the restrictions which fettered the development of commerce between Ireland and the other parts of the Empire, and would, had it been allowed to take effect, have removed one of the greatest causes of Irish misery in the following century.

(2) *The People and their Industrial Character.*

We have seen the terrible loss which Ireland suffered during the years of the Rebellion by the sword, famine, and plague, and later by emigration. This loss was to a small extent counterbalanced by the immigration of Americans,[1] and of English settlers, adventurers and soldiers. However, as the immigration was but a small matter compared with the emigration and other drains, we may conclude that the country was very sparsely peopled in 1660. A careful census was made in 1659, which has been brought to light in recent years.[2] The figures arrived at by an analysis of this census are as follows:—

Leinster	155,534
Ulster	103,923
Munster	153,282
Connaught	87,352
	500,091

In 1672 Petty estimated the population at 1,100,000, divided into 800,000 Catholics and 300,000 Protestants.[3] In 1676 he estimated the population as 1,200,000,[4] and in 1687 at 1,300,000.[5] Obviously Petty's estimate cannot be reconciled with that arrived at by the estimate of 1659, but it is quite impossible to say which is correct. On the one hand, it may be said that the census of 1659 was carefully made, and is, therefore, worthy of credence; in which case Petty must have considerably over-estimated

---
[1] Moran, *Persecutions of the Irish Catholics*, p. 332.
[2] Trans. R.I.A., vol. xxiv, part iii, p. 319.
[3] *Political Anatomy*, pp. 7-8.   [4] *Political Arithmetic*, p. 42.
[5] *A Treatise of Ireland.*

the population. This was the opinion of Newenham, who studied the question with great care.¹ On the other hand, Petty based his calculations on the hearth money returns, and it is notorious that the number of hearths in Ireland was greatly under-estimated. In this case Petty's figures were too small, and this was the opinion of Mr. G. P. Bushe, another careful investigator.² The only other returns which we have are an estimate made by Captain South in 1696, who, calculating from the poll tax return in three counties and the city of Dublin, reckoned the population at 1,034,102.³ A rough guess made in 1691 placed it at 1,200,000.⁴ It is quite impossible, with the material at our disposal, to come to any definite conclusion as to the numbers of the population in the period under review.

The economic value of the population was lessened by the great number of non-productive people it contained. Petty complained that one-fifth of the population were " casherers and faitneants,"⁵ and elsewhere refers to the abundance of " locusts and caterpillars of the Commonwealth."⁶ There were said to be too many clergymen, both Protestant and Catholic, in proportion to the population,⁷ and too many engaged in the administration of the law.⁸ Lawrence some years later complained of " the multitude of idle and unprofitable people, that, like drones in the hive, consume the honey others bring in."⁹ On the other hand, the industrial efficiency of the population was greatly strengthened by the immigration of many foreign Protestants from France and the Netherlands, who set up manufactures and introduced improved methods into the existing industries of the country.¹⁰

During the period from the Restoration to the Revolution, Catholics were practically unmolested. There were certainly no laws directed against their economic prosperity, except

---
[1] *Essay on Population*, 1805.   [2] Trans. R.I.A., vol. iii, p. 145.
[3] *Hull's Edition of Petty's Economic Works*, p. 142.
[4] *Remarks on Interest and Trade of England and Ireland*, London, 1691.
[5] *Political Anatomy*, p. 13.   [6] *Ib.*, p. 90.   [7] *Ib.*, pp. 16, 31.   [8] *Ib.*, pp. 37-8.
[9] *Ormond MSS.*, vol. iv, p. 39.
[10] On the great benefits which the industry of the British Isles derived from the immigration of foreigners in the latter half of the Seventeenth Century see Hewins, *English Trade and Finance*, pp. 107-8.

the regulations relating to their position in the corporate towns. Immediately after the Restoration all Catholics were given the right to carry on their trades fully and freely.¹ A year later some check was put on their admission to the corporate towns,² but there is abundant evidence that, as a matter of fact, the towns of Ireland were not closed to Catholics during succeeding years. The Act of Explanation provided that all houses in Corporations should remain in the hands of English and Protestant subjects,³ and the new regulations made for the government of the towns in 1667 prohibited any person to be mayor, sovereign, portreeve, burgomaster, bailiff, alderman, recorder, treasurer, sheriff, town clerk, common councilman, or master or warder of any guild, corporation or fraternity unless he took the oath of supremacy.⁴ In 1671 the towns were fully thrown open to Catholics, who were thenceforth to be permitted to reside and trade in them, but were still incapable of attaining to positions of dignity or importance in the corporations.⁵ Two years later, however, it was provided that no Catholic should inhabit a corporate town without a licence.⁶

It is probable, however, that the corporate towns were not closed to Catholics, who were allowed, without question, to carry on their occupations.⁷ The panic occasioned by the Titus Oates conspiracy aroused a temporary outbreak of bigotry against the Catholics of Ireland; orders were issued that no Catholics should be allowed to remain in any town save those who had resided there continuously for twelve months, and that no Catholics should be allowed to attend any fairs or market in any town.⁸ But it was one thing for the English Government to decree the removal of the Catholics from the towns, and another thing to get its orders carried out in Ireland, where it was realized that the presence of the Catholics was an economic necessity. "It is apparent,"

---

¹ C.S.P., Ire., 1660-2, p. 325.   ² Gale, *Corporate System of Ireland*, p. 74.
³ 17 & 18 Car. II., c. 2. sec. 36.   ⁴ Scully, *Penal Laws*, pp. 86-7.
⁵ C.S.P., Dom., 1671-2, p. 166.   ⁶ C.S.P., Dom., 1673, p. 559.
⁷ C.S.P., Dom., 1673-5, pp. 1, 157-9.   ⁸ *Ormond MSS.*, vol. v, p. 17.

wrote Ormond, "that trade cannot be carried on in towns, nor husbandry in the country, without some Popish merchants, and very many Popish tenants, unless a large plantation of English of all sorts could be sent us, for which we would be very glad to make a double return of Irish papists." Instructions were sent from England that a bill should be prepared incapacitating Catholics from sitting in either House of Parliament, or bearing any employment, civil or military, but apparently no such measure was ever seriously contemplated by the authorities at Dublin.[2]

During the reign of James II., all vestiges of penal distinction between Catholics and Protestants were removed. One of the first Acts passed by the Patriot Parliament was "an Act for Liberty of Conscience, and Repealing such Acts or Clauses in any Act of Parliament which are inconsistent with the same."[3] If the tolerant policy of this body had been imitated by its successors, most of the dreadful economic degradation of the Irish people in the eighteenth century would have been avoided.

Generally speaking, the Irish were during this period beginning to be accused of idleness by English writers, although the cause of this idleness was usually admitted fairly enough to be the lack of encouragement to work. "Their lazing," says Petty, "seems to me to proceed rather from want of Imployment and Encouragement to Work, than from the natural abundance of Flegm in their Bowels and Blood; for what need they to Work, who can content themselves with Potato's, whereof the Labour of one Man can feed forty; and with Milk, whereof one Cow will, in Summer time, give meat and drink enough for three Men, when they can every where gather Cockles, Oysters, Muscles, Crabs, etc., with Boats, Nets, Angles, or the Art of Fishing; can build an House in three days? And why should they desire to fare better, tho with more labour, when they are taught, that this way of living is

[1] *Ormond MSS.*, vol. v, p. 61.   [2] C.S.P., Dom., 1679-80, p. 311.
[3] Davis, *Patriot Parliament*, p. 51.

more like the Patriarchs of old, and the Saints of later times, by whose Prayers and Merits they are to be reliev'd, and whose Examples they are therefore to follow? And why should they breed more Cattel, since 'tis Penal to import them into England? Why should they raise more commodities, since there are not Merchants sufficiently Stock'd to take them of them, nor provided with other more pleasing foreign Commodities, to give in Exchange for them? And how should Merchants have Stock, since Trade is prohibited and fetter'd by the Statutes of England?"[1] Sir William Temple was also of opinion that the great abundance of provisions in the country, and the depleted condition of the population, were the reasons why the Irish did not exert themselves so energetically as they would do under different circumstances:—" In Ireland, by the largeness and plenty of their food, and the scarcity of people, all things necessary to life are so cheap that an industrious man by two days labour may gain enough to feed him the rest of the week."[2] " Fullness of bread is the cause of abundance of idleness in Ireland as well as in Sodom."[3] Too many religious holidays were said to be observed,[4] and these were regulated by Act of Parliament.[5]

Complaints of drunkenness are not very common during this period. The sale of drink was subjected to many regulations requiring licenses to be taken out and otherwise;[6] and, although Petty complains that too many people were employed in the manufacture and distribution of ale, he does not infer that too much was drunk. On the contrary, the reform which he advocated was the establishment of an arrangement by which the same amount of drink could be sold in a less wasteful manner.[7] Lawrence, in his *Interest of Ireland in its Trade and Wealth Stated*, complains of the prevalence of drunkenness, but too much reliance should not be placed on what

---

[1] *Political Anatomy*, pp. 99-100. [2] Temple. *Observations on the United Provinces*.
[3] Lawrence, *The Interest of Ireland in its Trade and Wealth Stated*, London, 1682.
[4] Petty. *Political Anatomy* p. 118; *A New Irish Prognostication*, London, 1689, p. 63. [5] 7 Will. III., c. 14. [6] 14 & 15 Car. II., c. 18; 17 & 18 Car. II., c. 19.
[7] *Political Anatomy*, pp. 13-14.

IN THE SEVENTEENTH CENTURY. 127

he says on this subject, on which he was obviously a fanatic. On the other hand, in a pamphlet of the period of the most scurrilous anti-Irish views,[1] it was asserted that drinking was not a national vice of the Irish.

(3) *The Land System.*

The whole ownership of land in Ireland was fundamentally changed by the Cromwellian war and the settlement which followed. Of the former landowners, one class were deprived of the whole of their estates, and were executed or banished; a second who offended less grievously against the Parliament forfeited a portion of their land, and were compensated for the remainder by an allotment of land elsewhere. " First of all, there were those who, not having joined the war before November 10th, 1642, had at any time served against the Parliament as colonel, or in any higher rank, or as governor of any castle or fort. They were to be banished for life, and their estates confiscated. But their wives and children were to receive lands to the value of one-third of their former estates wherever the Parliament should appoint. It is to be noted that to have borne arms against the *Parliament* was the ground for condemnation here. Secondly, those who, since November 10th, 1642, had at any time borne arms against the Parliament, but had not served as colonel, etc., were also to receive lands equal to one-third of their former estates wherever Parliament should appoint; but were not to be banished."[2] In this clause no religious distinction was made, but such a distinction was drawn in a later clause of the Act, which provided that all persons of the Popish religion who had resided in Ireland at any time between October 1st, 1641, and March 1st, 1650, and had not come under any of the previous clauses, were to lose one-third of their estates, and to get lands equal to the

[1] *A Brief Character of Ireland*, London, 1692.
[2] Butler, *Confiscation in Irish History*, pp. 127-8.

other two-thirds wherever the Parliament might appoint, unless they could prove *constant good affection to the commonwealth*. " And all other persons, *i.e.*, Protestants, who had been in Ireland at any time during the same period were to forfeit one-fifth of their estates unless they could show that they had been in actual service of the parliament, or had otherwise manifested good affection to its interests having opportunity to do the same. It is to be noticed that they were not required to prove *constant* good affection, and that they were to keep four-fifths of their actual estates.'" Here there appears a clear distinction between Catholic and Protestant landowners to the prejudice of the former—an anticipation of the Penal Laws which later were designed to oust the Catholics from all property in land. It is to be noticed that the presumption of guilt was against the landowner, and that the onus was on him to prove why his lands should not be forfeited. Outside Dublin, according to Sir William Petty, only 2,000 Catholic landowners, owning between them 40,000 acres, succeeded in proving " constant good affection " within the meaning of the Act; and with the exception of these 2,000, every Catholic landowner in Ireland lost his estates. Many of them lost everything, whereas many others were entitled to receive lands equivalent to one-third or two-thirds of their former estates in some other place appointed by the Government.[2]

In the earlier stages of the Cromwellian settlement there was no indication of the subsequent development of the transplantation system, which was first definitely decided on towards the end of the year 1653, when it was announced that Connacht and Clare had been selected as the region where all the Irish entitled to lands in lieu of their forfeited estates were to receive them. All the Irish were to remove beyond the Shannon before May 1st, 1654, and those found on the east side of the river after that date were condemned to death.[3] As a matter of fact,

---

[1] Butler, *op. cit.*, p. 129.   [2] Butler, *op. cit.*, p. 133.
[3] Butler, *op. cit.*, pp. 134-5.

the number who were actually made to transplant was not so great as is commonly supposed. In 1655 it was ordered that the only persons bound to transplant were landowners and those who had actually borne arms. Landowners included not only landlords but all those, their heirs, who might become landlords, mortgagees, and widows entitled to dower or jointure, and all tenants who held leases of lands for more than seven years.[1] "Thus the Cromwellian settlement meant the complete sweeping away of all Catholic landholders from all the counties east of the Shannon, and from two of the six counties west of that river. But it did not, as we have seen, involve the sweeping away of the land's inhabitants. These remained a despised but indispensable race, hewers of wood and drawers of water, for their conquerors." The big Catholic landlords were particularly hard hit by the settlement, as so few of them succeeded in escaping the provisions of the Act owing to their greater difficulty of doing so. "The act of Transplantation was simply an act of eviction, remarkable for the quantity of ground 'cleared,' and for the class of tenants upon whom notices to quit were served. The evicted were pre-eminently the upper tenants or landlords; they were turned out for (alleged) non-payment of their renders or services, *i.e.*, for breach of fealty."[2]

As regards Connacht, where the transplanted landlords and tenants settled down, the settlement contained many points which were commendable from the economic standpoint. For instance, although tenants were bound to transplant, they were not bound to go under their old landlord, but were usually given lands in fee simple directly dependent on the Crown. "They might sit down in Connacht as tenants under the State."[3] This, as we have seen, was in accordance with the general policy of seventeenth century statesmen, as indicated by Davies

---

[1] Prendergast, p. 130. It is impossible to arrive at a conclusion how many persons were actually transplanted under this Act; the question is discussed in Butler, *op. cit.*, pp. 149-51, where all the conflicting opinions are considered.
[2] Sigerson, *History of Irish Land Tenures*, p. 79.   [3] Sigerson, p. 80.

and Wentworth, and was not without benefit to the tenants themselves. "The valuable equivalents of lands given to the tenant-class under the Council Order at once raised their property status to an equality with, if it did not place many of them above, their former patrons and masters."[1] Indeed, the settlement of Connacht was in some ways a repetition of the composition of that province in the previous century, and it is the opinion of those who have studied the question most carefully that the establishment of this large body of small freeholders was one of the reasons why Connacht in after years continued to remain contented and prosperous during troubled periods.[2]

Although the settlement may have had this beneficial result, it was not planned with such good motives; the suppression of the old relationship between landlord and tenant being designed not to elevate the latter, but to degrade the former:—"the object was to degrade the evicted upper tenants or landlords to a lower condition. It was hoped they would be lowered to the rank of cultivators and earth tillers and peasants, by having to work for themselves."[3] Moreover, if something resembling a peasant proprietary was established in Connacht, the possibility of such a proprietary developing was abolished in the other three quarters of Ireland, from which the small freeholders, especially the Catholics, were almost universally expelled:—"One result of the confiscation, a result which the present generation is seeing reversed by means of costly machinery and after generations of discord, was the almost complete disappearance from the island of a peasant proprietary."[4] The lands assigned to the new settlers tended to get into fewer and fewer hands; the officers purchased large quantities of certificates from their soldiers at an undervalue; and also from the adventurers, most of whom were London citizens who did not want themselves to undertake the

---
[1] Hardinge. *Circumstances Attending the Civil War in Ireland*, Trans. R.I.A., vol. xxiv, ant. p. 401.
[2] Hardinge, *op. cit.*   [3] Sigerson, p. 80.   [4] Butler, *op. cit.*, p. 164.

hazardous enterprise of settling down as farmers in a strange country.¹ "The Cromwellian project," says Dr. Cunningham, "was devised so as to give the greatest possible shock to property; labourers were allowed to remain that they might till and herd for those to whom the lands were newly assigned, but the old proprietors were to go."²

There is no necessity here to go into the details of the Restoration settlements, full particulars of which may be found in many books. The question at issue amongst the contending parties was not whether the old proprietors should be restored to their former possessions, but which of a number of land grabbers should take their place. The last persons to be considered in the settlements were the old Irish, who were only to get their lands back after the Cromwellians in possession were reprised. It is the opinion of Mr. Butler that there were not more than 540 Catholics declared innocent—a figure which, if correct, means that the vast majority of the old landlords who had been dispossessed by Cromwell continued in that condition after the Acts of Settlement. The last hope of the old Irish being restored their possessions was taken away by the Act of Explanation;³ and in any case those who recovered their estates were in general the magnates, the lesser men who, of course, were far more numerous, being utterly ruined.⁴

The result of the juggling with the lands both before and after the Restoration, as it affected the proprietors, is admirably summarised in Mr. Butler's book on confiscation:—"In 1641 there were, at the lowest estimate, eight thousand Roman Catholic landowners in Ireland. Of these, all except twenty-six were deprived of their property by Cromwell. A certain number of the dispossessed received compensation west of the Shannon amounting to two-thirds or to one-third of their former holdings. We have seen that there is an official list

---
¹ Butler, *op. cit.*, p. 163.
² *Growth of English Industry and Commerce*, vol. ii., p. 366.
³ Butler, *op. cit.*, pp. 194-5. ⁴ Butler, *op. cit.*, p. 202.

extant from which it would appear that about two thousand persons were thus compensated. Now, if we come to the state of affairs after the execution of the Acts of Settlement and Explanation, we find that between 500 and 540 Catholics were restored as innocents; and that, when the status of the transplantees to Connaught was finally regularised in 1677, 580 persons received letters patent. If we add to these the nominees in the two Acts, and such of the letterees and ensignmen as ultimately recovered some portion of all their lands, we cannot allow a grand total of Catholic landowners for all Ireland under Charles II. of more than thirteen hundred at the outside."[1] The exact area actually forfeited is a matter of much discussion; though of much interest in itself, it is not of any importance in gauging the economic results of the Settlement. It is sufficient for this purpose to note that the bulk of the old proprietors were dispossessed and their places taken by unsympathetic strangers. The question of the area forfeited is fully and ably discussed in Butler on Confiscation,[2] where all the conflicting statements of contemporaries are considered, if not reconciled. Butler's own opinion is that, in 1685, only one-seventh or one-eighth of the total area of Ireland remained in the hands of Catholics.

Economically, the most important result of this change in the ownership of land was the feeling of insecurity which it created, and the consequent lowering of the value of land. In 1661 Petty stated that "lands in Ireland were worth but seven years' purchase: 1st—By reason of the frequent rebellions—in which, if you are conquered, all is lost, or if you conquer, yet you are subject to swarms of thieves and robbers—and the envy which precedent missions of England have against the statesmen; perpetuity is but forty years long, as within which time some ugly disturbance hath hitherto been almost ever since the first coming of the English thither. 2nd—The claims upon claims which each hath to the other's estates, and

[1] Butler, *op. cit.*, p. 197. [2] *Ibidem*, pp. 197-200.

the facility of making good any pretence whatsoever by the decision of some one or other of the many governors or ministers in power there. 3rd—By the paucity of inhabitants. . . . 4th—By absenteeism . . . . and 5th—The difficulty of executing justice, so many of those in power being themselves protected by officers and protecting others."[1] The ownership of land must have been insecure owing to the innumerable hordes of dispossessed proprietors and their dependents, the kerns or tories, who wandered through the country; there were actual instances of former owners evicting the new proprietors by force.[2] Richard Lawrence, in 1677, stated that one of the reasons why Ireland, notwithstanding its advantages for trade and wealth, was so backward in both, was "from the unsettledness of the minds of the people, the Irish envying rather than imitating the English in their modes, and the English, jealous of disturbance from the Irish, discouraged from industry in improving."[3]

The Irish who did not emigrate or transplant to Connacht with their landlords were universally taken on as tenants by the new settlers, who generally had no knowledge themselves of agriculture, and had no English tenants to place on their farms. "The officers, immediately upon obtaining a lease or custodium of their estate, took the Irish as tenants for want of English; for in a country where lands were to be had for the asking no one would come from a better country to a worse to labour as a servant on another man's land when he might till or pasture his own. As the impossibility of getting English tenants grew more evident, and the urgent want of tillage increased, the officers in Cork, Limerick, Kerry, and various other counties got general orders giving dispensations from the necessity of planting with English tenants."[4] Here we see a repetition of what had happened in previous plantations; the Irish crept in as tenants to the English, in spite of the law. The inevitable amalgamation of races again commenced, although they did not

[1] Petty's *Treatise of Taxes*, Hull's Edition, pp. 7, 46. [2] C.S.P., Dom., 1672-3, p. 580.
[3] *Ormond MSS.*, vol. iv., p. 390. [4] Prendergast, p. 266.

merge so completely as former generations of planters and Irish had done, owing to the difference of religion, which now prevented the intermarriages from being as frequent as before.

Moreover, the tenants in this period continued to insist on their tenant right, and therefore possessed the land on a more satisfactory basis than did the degraded tenants of the eighteenth century, from whom, except in Ulster, all customary rights were wrested. Dr. Sigerson points out that it was no part of the intention of the Cromwellian governors or settlers to abrogate the old customary law of landlord and tenant in Ireland, which was impliedly recognised by the provisions which allowed the transplanted upper tenants or landlords to go back in order to reap and carry off their way-going crops, charging, with various percentages, the new landlord according to the custom and the locality.[1] The existence of customary rights is further recognised in the Acts which substituted socage tenures for knights' service, and which declared that alienation fines due by the particular customs of principal manors should not be abolished.[2] It is probable that leases in this period were no more common than in the earlier part of the century, and that tenants remained for several years on the same holding relying on the custom of the country, enforced by the sanction of the tories and other agrarian agitators who abounded. The Statute of 1695, invalidating all leases for more than three years which were not in writing, is another recognition of the existence of the Irish custom of the country, which later disappeared everywhere except in Ulster. We are, therefore, justified in concluding that the Irish tenants in the latter half of the seventeenth century were protected against their landlords more effectively than were those of the eighteenth, and there is no evidence that the great evils which characterised the Irish land system at a later period had yet begun to appear. The principal agrarian suffering of this period was

[1] Sigerson, p. 83; Prendergast, pp. 35-7.   [2] Sigerson, pp. 93-4.

experienced, not by the tenants, but by the landlords, who were liable to lose their estates when a hostile party came into power in the Government. Above all, it must be remembered that there was no religious disability to hold land, and that in this respect Catholics and Protestants were on an absolute equality. There was, therefore, nothing to discourage the Catholics, who, of course, represented the vast majority of the population, from improving their lands, as they might confidently expect to reap the benefit of their improvements, being entitled to long leases, which they were not forbidden to take, or, as was probably more common in practice, being protected by the custom of the country, which, as we have seen, was generally recognised.

One of the worst effects of the Cromwellian and Restoration settlements was that they caused an increase of absenteeism, which first began in this period to assume something of the magnitude which it afterwards attained in the eighteenth century. This, of course, was owing to the great juggling with estates which took place, whereby large Irish interests passed into the hands of English proprietors who had no desire to live in Ireland. In 1662 Petty noticed that lands were worth fewer years purchase in Ireland on account of the change:—" but a great part of the estates both real and personal are owned by absentees and such as draw over the profits raised out of Ireland refunding nothing, so that Ireland exporting more than it imports doth yet grow poorer through charges."[1] Ten years later Petty again noticed the existence of this evil, and calculated that the owners of close on a quarter of the real and personal estate of Ireland lived in England;[2] and Sir William Temple drew attention to the growth of absenteeism about the same time.[3] "There is nothing," wrote Ormond, in 1677, "the considering part of this people do more apprehend should prove their ruin in the end, or at the best hindering them

---

[1] *Treatise on Taxes*, p. 28. [2] *Political Anatomy*, p. 72.
[3] Temple, *Essay on the Trade of Ireland*, 1673.

from reaping of the fruit of their industry than the transmission of their money into England.'"¹ In 1682 the amount remitted annually to England to absentee landowners was calculated at £157,465.² In 1691 the amount of money transmitted to absentees was estimated at £136,017.³

(4) *The Cottiers.*

The cottiers remained a constant element in the Irish population throughout the many changes and revolutions which occurred in the middle of the seventeenth century. Of course, great numbers perished from famine and plague during the wars, but those who survived were, on the whole, left undisturbed in their former condition. The Acts of Transplantation had always excepted from their provisions all ploughmen, husbandmen, labourers, artificers, and others of the inferior sort, if they were not possessed of goods to the value of ten pounds, and if they had not come under those classes excepted from pardon. Of course, a great many of them had, no doubt, borne arms, and were either executed or banished, but on the whole it may be said that there was no clean sweep made of the labouring class as there was of the landlord class.⁴ " The tempest that devastated the castle swept over the cabins, the cultivators bent to the storm, but sprang up when its strength had passed."⁵ Indeed, the new land settlement was probably advantageous to the labouring class:—" The great change in the condition of the farming class brought about by the transplantation may not have been intended, though it was inevitable— nor did the beneficial effects end here; they extended themselves to the humbler classes by the increases made to the number and wealth of their employers, and the consequent extra demand for and price of labour."⁶

---

¹ *Ormond MSS.*, vol. iv., p. 55.
² Laurence, *The Interest of Ireland in its Trade and Wealth Stated*, London, 1682.
³ *Remarks on the State of Trade of England and of Ireland*, London, 1691.
⁴ The extent of this exception is discussed in Gardiner, *History of the Commonwealth*, vol. iii, p. 202.   ⁵ Sigerson, p. 87.
⁶ Hardinge, *Circumstances Attending the Civil Wars in Ireland*, Trans. R.I.A., vol. xxiv., ant. p. 401.

IN THE SEVENTEENTH CENTURY.

During the latter half of the seventeenth century by far the greater part of the population of Ireland belonged to this cottier class. Petty estimated that 16,000 people had houses with more than one chimney, 24,000 had houses with one chimney, and 160,000 had houses without any chimney at all.[1] Probably this lower class tended to increase more rapidly than at an earlier period owing to the very general introduction of potatoes as a staple food, and the ease with which a bare sustenance could thereby be obtained. Indeed, the transition of the Irish cottier from a person living on the produce of his cattle to one living mainly on potatoes probably marks an important turning point in Irish history. In 1655 it was said that the Irish lived on " the roots and fruit of their gardens "—the roots probably being potatoes,[2] and potatoes were mentioned as a staple food in 1684.[3]

The condition of these peasants seems to have neither improved or disimproved since the earlier part of the century. "Their houses are not to be gone but groped into, they making their doors as low and as little as they can and their ceilings being as low as a man's head."[4] An attempt was made to improve the condition of the peasantry in 1666 by a statute which provided that no cabin should be let unless an acre of land were let with it,[5] but this provision was probably not carried out, as we know that the other sections dealing with the sowing of flax and hemp quickly became a dead letter. The following account of the cottiers was given by a traveller in 1672 :— " The cots are generally built on the side of a hill not to be discerned till you just come upon them. The cottage is usually raised three feet from the eaves to the ground on the one side, and on the other side hath a rock for a wall to save charges. . . . The hearth is placed in the middle of the house, and their fuel is made of earth and cow dung dried in the sun. The smoke goes through no

[1] Petty, *Political Anatomy*, p. 9.
[2] Gookin, *Great Case of Transplantation Discussed*, p. 15 ; see also Cole, *Adam in Eden*, London, 1657 ; and a Paper by W. R. Wilde, in Proc. R.I.A., vol. vi., p. 356.
[3] *Chorographic Account of the Southern part of the County of Wexford*, Kilk. Arch. Jl., n.s. vol. ii., p. 466.
[4] Col. Ed. Cook to Lord Bruce, *Somerset MSS.*, p. 169.   [5] 17 & 18 Car. II., chap. 9.

particular place, but breaks through every part between the rods and wattles of which they make the door sides and roof of the house, which commonly is no bigger than an overgrown pigstye to which they have two doors, one always shut, on that side where the wind blows. Their general food is a thin oatcake which they bake upon a broad flatstone made hot, a little sheepsmilk cheese, or goat's milk, boiled leeks and some roots."[1]

Of course, the fullest account of the cottiers' condition which we have is to be found in Petty's *Political Anatomy*, where it is stated that six out of eight of all the Irish " live in a brutish nasty condition as in cabins with neither chimney, door, stair nor window, and feed chiefly upon milk and potatoes."[2] The following is his description of the peasantry :—" That is to say Men live in such Cottages as themselves can make in 3 or 4 days; eat such food (Tobacco excepted) as they buy not from others; wear such Cloaths as the Wooll of their own Sheep, spun into Yarn by themselves, doth make; their Shoes, called Brogues, are but quarter so much worth as a pair of English Shoes; nor of more than quarter in real use and value. A Hat costs 20d., a Pair of Stockins 6d., but a good Shirt near 3s. The Taylors work of a Doublet, Breeches and Coat, about 2s. 6d. In brief, the victuals of a man, his wife, three children, and servant, resolved into Money, may be estimated at 3s. 6d. per week, or 1d. per diem. The Cloaths of a Man, 30s. per Ann. of Children under 16, one with another 15s. the House not worth 5s. the Building; Fuel costs nothing but fetching. So as the whole annual expence of such a Family, consisting of 6 in number, seems to be about 52 shillings per Ann. each head one with another. . . . The Housing of 160,000 Families, is, as hath been often said, very wretched. But their Cloathing far better than that of the French Peasants, or the poor of most other Countreys; which advantage they have from their Wooll, whereof 12 Sheep furnisheth

---

[1] *A Tour in Ireland in* 1672-4, Jl. of Cork Arch. Soc.. vol. x., p. 96.
[2] Petty, *Political Anatomy*, p. 27.

a competency to one of these Families. Which Wool, and the Cloth made of it, doth cost these poor people no less than £50,000 per Ann. for the dying it; a trade exercised by the Women of the Country. Madder, Allum, and Indico, are imported, but the other dying stuffs they find nearer home, a certain Mud taken out of the Bogs serving them for Copperas, the Rind of several Trees, and Saw-dust, for Galls; as for wild and green Weeds, they find enough, as also of Rhamnus-Berries.

"The Diet of these people is milk, sweet and sower, thick and thin, which also is their drink in Summer-time, in Winter Small-Beer or Water. But Tobacco taken in short Pipes seldom burnt seems the pleasure of their Lives, together with Sneezing; Insomuch that two-sevenths of their expense in Food, is Tobacco. Their food is Bread in Cakes, whereof a Penny serves a Week for each; Potatoes from August till May, Muscles, Cockles and Oysters, near the Sea; Eggs and Butter made very rancid, by keeping in Bogs. As for Flesh, they seldom eat it, notwithstanding the great plenty thereof, unless it be of the smaller Animals, because it is inconvenient for one of these Families to kill a Beef, which they have no convenience to save. So as 'tis easier for them to have a Hen or Rabbit, than a piece of Beef of equal substance.

"Their Fewel is Turf in most places; and of late, even where Wood is most plentiful, and to be had for nothing, the cutting and carriage of the Turf being more easy than that of Wood. But to return from whence I digressed, I may say, That the Trade of Ireland among 19/22 parts of the whole people, is little, or nothing, excepting for the Tobacco above-mentioned, estimated worth about £50,000, for as much as they do not need any Foreign Commodities, nor scarce anything made out of their own Village. Nor is above one-fifth part of their Expence other than what their own Family produceth, which Condition and state of living cannot beget Trade.'"

[1] Petty, *Political Anatomy*, pp. 76-82.

Living was extremely cheap in Ireland at this time; a quarter of mutton could be obtained for eightpence, a salmon for tenpence, and twenty eggs for a penny.[1] In spite of their poverty, the cottiers were much given to mutual hospitality; the following account of a gathering of cottiers is taken from a pamphlet written in 1689:—
"The manner of their sitting in this great feasting is this; stools nor table they have none, but a good bundle of straw strewed about the floor, they sit themselves down one by the other; another bundle of straw doth serve them to set on their dishes; perhaps if it be in the time of summer or where the place will afford it then in the stead of straw they use green rushes. Victuals they have plenty, beef, mutton, pork, hens, rabbits, and altogether served in a great wooden platter; aqua vitae they must have. . . . They have no bread but oats."[2]

Stevens, writing in 1690, said:—"The meaner people content themselves with little bread but instead thereof eat potatoes, which with sour milk is the chief part of their diet, their drink for the most part water, sometimes coloured with milk; beer or ale they seldom taste unless they sell something considerable in a market town. They all smoke, women as well as men, and a pipe an inch long serves the whole family several years, and though never so black or fowl is never suffered to be burnt. Seven or eight will gather to the smoking of a pipe, and each taking two or three whiffs gives it to his neighbour, commonly holding his mouth full of smoke till the pipe comes about to him again. They are also much given to taking of snuff. Very little clothing serves them, and as for shoes and stockings much less. They wear brogues being quite plain without so much as one lift of a heel, and are all sowed with thongs, and the leather not curried, so that in wearing it grows hard as a board, and therefore many always keep them wet, but the wiser that can afford it grease them often and that makes them supple. In the

[1] *Tour in Ireland in* 1672-4, Jl. of Cork Arch. Soc., vol. x., p. 89; and see *The Present State of Ireland*, London, 1673.
[2] *A New Irish Prognostication*, London, 1689, p. 40.

better sort of cabins there is commonly one flock bed, seldom more, feathers being too costly; this serves the man and his wife, the rest all lie on straw, some with one sheet and blanket, others only their clothes and blanket to cover them. The cabins have seldom any floor but the earth, or rarely so much as a loft, some have windows, others none. They say it is of late years that chimneys are used, yet the house is never free from smoke. That they have no locks to their door is not because there are not thieves but because there is nothing to steal. Poverty with neatness seems somewhat the more tolerable, but here nastiness is in perfection, if perfection can be in vice, and the great cause of it, laziness, is most predominant. It is a great happiness that the country produces no venomous creature, but it were much happier in my opinion did it produce no vermin."[1]

"Their food is mostly milk and potatoes, their clothing coarse bandrel cloth and linen, both of their own make; a pot of gruel; a griddle whereon to bake their bread, a little salt, snuff, and iron for their ploughs being almost all they trouble their shopkeeper or merchant for. A little hut or cabin to live in is all that the poverty of this sort hope or have ambition for."[2] "Their dwellings or cabins—an English cow-house hath more architecture far; the Lord Mayor's dogs kennel is a palace compared to them; the walls are made of mere mud mixed with a little wet straw. For beds instead of feathers and flocks they use rushes and straw."[3] "Such of them as live in the plain and fertile parts of the country are generally slaves to the English or to their Irish landlords, and live by their daily labour working for threepence or fourpence a day and their dinner, their stock is generally a cow or two, some goats, and perhaps six or eight small Irish sheep which they clip twice a year and convert their wool into coarse frieze to cover their nakedness; when the lambs

---

[1] *Journal of John Stevens*, 1689-1691, edited by Dr. R. Murray, Oxford, 1912, pp. 139-40.
[2] *Remarks on Trade and Interest of England and Ireland*, London, 1691; and see *A Brief Character of Ireland*, London, 1692.
[3] *A Brief Character of Ireland*, London, 1692.

of their few sheep do fall they preserve some small part of them to keep up their number and those they half starve for lucre of the milk, the rest they sell in the markets for sixpence or eightpence a piece."¹

(5) *The State of Agriculture.*

The conditions of the Cromwellian Settlement were not favourable to the progress of agriculture. In the first place, the shock which had been given to security in the country did not encourage the new tenants to lay out their capital on the improvement of their lands, as they had no guarantee that the lands would not in turn be confiscated in a few years; and in the second place the planters were not themselves skilful in agriculture, being almost all either soldiers or citizens of London. Indeed, during the first few years of the plantation the planters were dependent for their food on the agricultural labours of the Irish, who attracted the attention of the planters by their skilful husbandry. "There are few of the commonalty but are skilful in husbandry, and more exact than any English in the husbandry proper to that country—there are few of the women but are skilful in dressing hemp and flax, and making of linen and woollen cloth—to every hundred men there are five or six masons and carpenters at least of that nation, and these more handy and ready in building ordinary houses and much more prudent in supplying the defects of instruments and materials than English artificers."²

During the whole of this period the old custom of ploughing by the tail continued. Temple was in favour of the strict enforcement of the statutes "against that barbarous custom of ploughing by the tail, and upon absolute forfeitures instead of penalties, the constant and easy composition whereof have proved rather a letting than a forbidding of it."³ The custom was observed to continue in 1681.⁴ The other objectionable custom of

[1] *A Discourse Concerning Ireland*, London, 1697-8.
[2] Gookin, *The Great Case of Transplantation Discussed*, London, 1655.
[3] *Essay on the Trade of Ireland*, 1673. [4] *Dinely's Tour*.

burning corn in the straw also continued:—" In the remote part of Ireland instead of threshing their corn they use to burn them in the straw and then winnowing them in the wind from the burnt ashes make them into meal."[1] A pamphlet of 1692 gives a very poor account of the condition of husbandry in Ireland; ploughing by the tail and burning corn in the straw were widely practised and the corn grown was inferior. " Enclosures are very rare amongst them and then no better fenced than an old midwife's toothless gums .... as for the arable ground it lies almost as much neglected and unmanured as the sandy deserts of Arabia."[2] In the early part of the century a great deal of land had been reclaimed by the drainage of bogs, which, according to Boate, was very skilfully done, but this altogether stopped on the outbreak of hostilities, and much of the land which was in process of reclamation again became waste.[3] However, when the country again became more settled, the process of reclamation was renewed with good results. " Much good ground is being gained, some fit for corn, some for pasture and some for meadow."[4]

The quality of Irish products was not as high as that of English. It was said in 1677 that Irish corn was fit only for home use, " being by reason of the climate not so large, firm, and dry a grain as should be fit for transplantation,"[5] and Stevens, in his Journal, written in 1690, complained of the inferior quality of Irish barley.[6] Petty gave the following as the amount of seed required and the return for each acre of Irish land on the average:—

Crop.	Seed.	Produce.
	Bushels.	Bushels.
Wheat	4	16 to 36
Rye	4	20 ,, 40
Bean	6	20 ,, 48
Oats	6	16 ,, 32
Barley	4	20 ,, 40
Pease	4	12 ,, 18

[1] *A New Irish Prognostication*, London, 1689.
[2] *A Brief Character of Ireland*, London, 1692.
[3] Boate, *Ireland's Natural History*, 1652, ch. 14.
[4] *Bath MSS.*, vol. ii., p. 126.
[5] *Letter from a Gentleman in Ireland to his Brother in England*, London, 1677.
[6] *The Journal of John Stevens*, Oxford, 1912, p. 163.

One of the effects of the Cattle Act, as we shall see, was greatly to increase the area of land under sheep, and incidentally to increase the quality of the animals bred, which Temple thought were fully equal to English sheep.[1] The Irish horse, on the other hand, did not improve at this time. " Horses are a drug, but could be made a commodity. The present defects in them are breeding without choice of stallions either in shape or size, and trusting so far to the gentleness of the climate as to winter them abroad, without ever handling colts until they are four years old; this both checks the common breeds and gives them an incurable shyness. .... In the studs of persons of quality in Ireland, where care is taken and cost is not spared, we see horses bred of excellent shape and vigour and size so as to reach great prices at home, and encourage strangers to find the market here, amongst whom I met with one this summer that came over on that errand and bought about 20 horses to carry over into the French Army from twenty to three score pounds apiece at the first hand."[2]

(6) *Pasture and Tillage.*

There cannot have been any extensive tillage in Ireland at the time of the Restoration, owing to the extreme sparseness of the population which had survived the Civil Wars, and to the inevitable shortage of labour which must have been caused by emigration and transplantation. Tillage was also said to be discouraged in this period by the growing custom of importing brandy instead of manufacturing whiskey in Ireland,[3] and by the proclamations issued by the Commonwealth against the sale of beer and ale.[4] The great amount of land under pasture may be gathered from the plenteousness of all sorts of cattle and sheep; in 1661 beef in Dublin was not worth a penny per pound;[5] and in 1667 the price of a

---
[1] *Essay on Trade of Ireland*, 1673.   [2] Temple, *Essay on Trade of Ireland*, 1673.
[3] C.S.P., Ire., 1647-60, p. 326.   [4] *Egmont MSS.*, p. 530.
[5] C.S.P., Ire., 1666-69, p. 6.

cow was only 10s.[1] In 1672 Petty estimated that not more than 500,000 acres of land were devoted to tillage, whereas seven million acres were given to pasture, and that the former gave employment to 100,000 people and the latter to only 120,000.[2] In 1685 the pasture lands of Ireland were said to be " proportionable to the bigness of the Kingdom far larger than the pasture land of England."[3] " We ploughed no more than might serve us yearly from hand to mouth for our own spending."[4]

But although, as we see, pasture outweighed tillage, there was nothing like the serious lack of tillage which characterized the eighteenth century, and Ireland throughout this period continued to be a corn exporting country, although not on a large scale. " We do not export much corn, as the damp climate renders our corn of little value; we have enough such as it is of all sorts for ourselves and some to spare our neighbours."[5] The exports of corn from Ireland in 1665 were as follows[6]:—

Corn.	To Foreign Parts.	To England.	Total.
Wheat, qrs.	135	740	875
Beans, qrs.	447	44	491
Peas, qrs.	10	15	25
Barley, qrs.	150	197	347
Malt, qrs.	355	989	1344
Bere, tuns	54¼	4¼	58½
Oats, qrs.	1251	1741	2992
Oatmeal, qrs.	476½	1084½	1561
Rapeseed, qrs.	475	171	646

In 1667 a proclamation was issued for setting up granaries in various parts of the country, as the first half of two subsidies then due were to be paid in wheat or oatmeal; this would seem to show that corn was plentiful in Ireland at the time.[7] In 1669 a proclamation was issued in England against the importation of foreign corn, in which Irish corn was at first included; but the

[1] C.S.P., Ire., 1666-9, p. 429. [2] Petty, *Political Anatomy*. pp. 12 and 54.
[3] *A Summary of Certain Papers*, Ed. by Wall, London, 1685. p. 51.
[4] *Letter from a Gentleman in Ireland to his Brother in England Relating to Trade*, London, 1677.
[5] C.S.P., Ire., 1662-6, p. 693. [6] C.S.P., Ire., 1662-6, p. 693. [7] C.S.P., Ire., 1666-9, p. 315.

hardship felt in Ireland as a result of this was so great that it was altered so as to except Irish corn from its provision.[1] In 1670 we read that corn in Ireland was cheap and plentiful;[2] but four years later the export of all sorts of corn was prohibited owing to the scarcity in the country.[3] In 1684 the principal commodity produced in Connacht was cattle of all sorts, " but it yields as much corn of wheat, barley, oats and rye as is enough to sustain the inhabitants and furnish the markets beside."[4] On King William's arrival at Carrickfergus he found the country full of corn,[5] and an English traveller in the same year was struck " by the valleys crowned with cornfields."[6]

Of course, the fact that there was enough corn grown in Ireland to meet the demands of the Irish consumer does not mean that the quantity was very great, because it must be remembered that the peasants, on the whole, consumed very little corn of any kind, but lived almost entirely on the products of their cattle. " None but the best sort or the inhabitants of great towns eat wheat or bread baked in an oven or ground in a mill; the meaner people content themselves with little bread, but instead thereof eat potatoes which with sour milk is the chief article of diet."[7] However, the condition of Ireland in the latter part of the seventeenth century compares favourably with its condition one hundred years later, when the greater part of the corn consumed in Ireland had to be imported from abroad, and when the exports of corn of all sorts had sunk to practically nothing. One might think one of the effects of the Cattle Acts would have been a great extension of the amount of tillage in Ireland, but this does not seem to have been so. The land was rather converted from cattle pasture to sheep pasture.

[1] C.S.P., Ire., 1666-9., p. 772.   [2] C.S.P., Ire., 1669-70, p. 258.
[3] C.S.P., Dom., 1673-5, p. 251.
[4] *Chorographical Description of Iar Connaught*, Irish Archæological Society, vol. ix, p. 15.   [5] *Ulster Journal of Archæology*, vol. i, p. 134.
[6] Eachard, *An Exact Description of Ireland*, London, 1691.
[7] *Journal of John Stevens*, Oxford, 1912, p. 139.

(7) *The Woods.*

The woods, as we have seen, were greatly wasted in the Rebellion, and one of the first measures of the Commonwealth Government was to prevent them from being altogether destroyed. The adventurers who had been granted lands, however, were profligate of their timber, and wasted it to such an extent that the Government forced them to exchange part of their woods for agricultural lands of equal value elsewhere.[1] In 1656 the exportation of timber was prohibited.[2] After the Restoration the lack of timber was greatly felt; the rebuilding of Dublin was delayed by the inability of the builders to get wood.[3] The woods were denuded and still further wasted largely on account of the great quantities of timber which were exported; large numbers of barrel staves were annually sent to France;[4] and even as late as 1685 timber still continued to be exported.[5] In 1691 timber was exported for the English navy.[6]

Although the iron works were, as we shall see, not worked so extensively as at an earlier period, they still had an appreciable effect in lessening the timber supply of the country.[7] Another practice which played havoc with the oak woods was the barking of live trees.[8] Of course, the insecurity of tenure which followed on the Acts of Settlement also tempted landowners to waste their timber; this was a feature which was very conspicuous during the whole of the seventeenth century.

" All the large timber has been spoilt by the brogue makers, who have stripped off the bark three or four feet from the root, which has caused the trees to decay as they stand; and what they leave is cut down and converted into barrel staves, so that between the brogue makers and the barrel makers most of the best woods of Ireland are destroyed."[9] " The land is so exceedingly denuded

[1] Dunlop, *Ireland Under the Commonwealth*; C.S.P., Ire., 1647-60, p. 851.
[2] *Ibid.*, p. 601. [3] C.S.P., Ire., 1663-6, p. 550. [4] C.S.P., Ire., 1660-2, pp. 166 & 412.
[5] Petty, *Treatise on Ireland*. [6] C.S.P., Dom., 1690-1, p. 220.
[7] Temple, *Essay on the Trade of Ireland*, 1673. [8] *Ibid.*
[9] C.S.P., Dom., 1671, p. 184.

148   THE ECONOMIC HISTORY OF IRELAND

and barren of trees that for ten or twenty miles together you could not see a bush to tie your horse to.'"[1]

Occasional efforts were made to prevent the complete destruction of the forests; the export of timber was from time to time prohibited for a short period, and the King forbade the sale of the Royal woods for any purpose except shipbuilding.[2] The process was also probably arrested to some extent by the substitution of turf for wood for firing,[3] but this substitution is in itself evidence that the woods were becoming very much wasted. In 1698 an Act was passed providing that all persons having freehold estates of ten pounds and upwards and tenants for years having eleven years unexpired should plant a certain number of trees in proportion to their holding,[4] but this Act was not a success.[5]

(8) *The Mines.*

As we have seen, all the mines of Ireland suffered greatly during the ravages of the Rebellion, and they did not quickly recover their former prosperity. The exports of metal after the Restoration were much smaller than they had been before the Rebellion[6] :—

	EXPORTS.		
	1641	1665	1669
Iron	Tons. 778	Tons. 56	Tons. 28
Lead	201½	—	1½

In 1660 a new patent was granted to Sir George Hamilton to work the silver mine in County Tipperary,[7] but the mine does not seem to have attained any output.[8] Indeed this grant was not only not acted on, but proved a discouragement to others.[9] In 1682 there was no silver, but only lead

---

[1] *A Tour in Ireland in* 1672-4, Jnl. of Cork Arch. Soc., vol. x, p. 90.
[2] C.S.P., Ire., 1660-2, p. 429.   [3] Dinely's *Journal.*   [4] 10 Wm. III., ch. 12.
[5] O'Brien, *Economic History of Ireland in the Eighteenth Century*, p. 154 ; on the wastage of timber during this period see Litton Falkner, *Illustrations of Irish History*, pp. 148-9 ; Journal of the Dublin Statistical Society, vol. xi, p. 162.
[6] C.S.P., Ire., 1669-70, p. 55.   [7] C.S.P., Ire., 1660-2, p. 153.
[8] A further grant of all the gold, silver and lead mines was made in 1671, C.S.P., Dom., 1671-2, p. 368.   [9] C.S.P., Dom., 1695, p. 207.

obtained in the mines, which were frequently rendered idle through the shortage of water in the vicinity.[1] Shortly before the Revolution an invention was introduced into Ireland for smelting and refining lead ore with pit or sea coal.[2]

In 1661 some Englishmen set up ironworks near Enniscorthy and brought over many English workpeople and their families,[3] but the progress of the ironworks in Ireland for the next twenty years was very slow. In 1665 there were complaints of the great difficulty of disposing of the products of the works owing to the large importations of iron from Spain and Sweden which depressed the price.[4] Petty says that there were more than ten iron furnaces and about twenty forges and bloomeries at work and one lead work, "though many in view which the pretended patents of them have hindered the working of. There is also a place in Kerry for one alum work attempted, but not fully proceeded on."[5] Petty further stated that there were "employed about making a thousand tons of iron two thousand men and women, and fifteen thousand smiths and their wives."[6] In 1673 it was said that there were no ironworks except at Mountrath and near the town of Wexford.[7] In 1677 there were still complaints of the backwardness of mining in Ireland. "I do not wonder that you make no considerable advantage by the concession of the grant of the mines of Ireland, neither can you ever expect you shall do so so long as you are not in a position to work them yourself, and that there are but few men in this nation so forward as to work them for you, this country being so inexperienced in matters of that kind."[8] "There is more iron," wrote Petty in 1687, "exported out of Ireland than imported into it, and consequently all the quantity of iron used in Ireland is made there."[9] The only reference we can find

---

[1] Dinely's *Jl.*, Kilkenny Archæological Jl., vol. v, n.s. p. 272. An interesting picture representing the mines at that date is to be found in the Journal.
[2] C.S.P., Dom., 1689-90, p. 563.   [3] C.S.P., Ire., 1660-2, p. 474.
[4] C.S.P., Ire., 1662-6, p. 540.   Some doubt is thrown on this by another State Paper of the same year, *Ib.*, p. 502.   [5] *Political Anatomy*, p. 111.   [6] *Ibid.*, p. 12.
[7] *The Present State of Ireland*, London, 1673.
[8] *Anonymous Letter to Sir George Hamilton*, Ormond MSS., vol. iv, p. 131.
[9] Petty, *Treatise of Ireland*.

to a copper mine in this period is to one worked by Sir William Petty himself at Kenmare.¹

The coal mines at Idough were worked during this period, but apparently only on a very small scale.²

(9) *Trade*.

During the period from the Restoration to the Revolution, Irish trade increased satisfactorily. This was largely the result of the public spirit of the Duke of Ormond, who did all he could to benefit the industries and commerce of the country. When the tonnage and poundage bill was being drafted in 1662 a special committee of the House of Commons was appointed to confer with the merchants of Dublin, so that the bill should not contain any clause detrimental to the interests of the Irish trading community;³ and about the same time a Committee of Trade was appointed to "debate and consider of those ways and means by which the trade of this Kingdom may be best and most effectually advanced."⁴ It was directed that notices should be put up in public places inviting all those who had "anything to propound for the encouragement of trade to repair and offer the same unto the said Committee."⁵ This Committee suggested many measures, which, if they had been adopted, would have done much to enrich Ireland. It made many proposals for the encouragement of the linen manufacture;⁶ recommended public registries of land;⁷ sought to have it ordered that some part of the English fleet should be regularly victualled in Ireland;⁸ and did its best to encourage the growth of native shipping.⁹ A report from the Council of Trade on "how the wealth of the Kingdom in general, and the money thereof in particular may be increased," was prepared by Sir William Petty, and forms the basis of his Political Anatomy. It is interesting to notice that the council

---

¹ Fitzmaurice, *Life of Petty*, p. 149. ² *Ormond MSS.*, vol. iii, p. 180.
³ *I.C.J.*, vol. i, p. 531. ⁴ *I.C.J.*, vol. i, p. 405. ⁵ *I.C.J.*, vol. i, p. 547.
⁶ *I.C.J.*, vol. i, p. 571. ⁷ *I.C.J.*, vol. i, p. 650. ⁸ C.S.P., Ire., 1666-9, p. 629.
⁹ C.S.P., Ire., 1666-9, p. 609.

was in favour of the adoption of a non-importation agreement. But it was easier to make suggestions than to get them carried out, and many of the excellent proposals of this committee never became realities, "as the Council of Trade was their nursery, so the Council table was their sepulchre."[1] Ormond encouraged numbers of skilled foreign artisans to settle in Ireland, and thus laid the foundation of many flourishing industries.[2] All the former inhabitants of the towns who had been expelled during the Rebellion, and who had practised useful trades, were allowed to return to their homes;[3] and the old rights of the corporation to prevent workmen from setting up in the towns unless they had served their apprenticeship there were restrained.[4] New trading corporations, such as the Merchant Adventurers of Munster, were given charters.[5]

On the other hand, during this period the principle that Ireland's prosperity must take a definitely secondary place to England's was for the first time systematically acted upon. As early as 1660, Charles II., in his draft instructions to Lord Robartes, whom he intended to send as Lord Lieutenant to Ireland, gave the following directions: "You shall in all things endeavour to advance and improve the trade of our Kingdom so far as it shall not be a prejudice to this our Kingdom (England), which we mean shall not be wronged however much that our other Kingdom might be concerned in it."[6] Sir William Temple also recognised the same principle: "As in the nature of its (Ireland's) government, so in the very improvement of its trade and riches it ought to be considered not only in its own proper interest, but likewise in its relation to England, to which it is subordinate, and upon whose weal in the main that of this Kingdom depends; and therefore a regard must be had to those points wherein

---

[1] Lawrence, *The Interest of Ireland in its Trade and Wealth Stated*, London, 1682.
[2] C.S.P., Ire.. 1660-2, p. 415. This is more fully dealt with when treating of the particular industries thus founded or encouraged.
[3] C.S.P., Ire., 1660-2, p. 339. [4] C.S.P., Ire., 1660-2. p. 301.
[5] C.S.P., Ire., 1669-70, p. 171. [6] C.S.P., Ire., 1660-2, p. 16.

the trade of Ireland comes to interfere with any main branches of the trade of England; in which case the encouragement of such trade ought to be either declined or moderated, and so give way to the interest of trade in England, upon the health and vigour whereof the strength, riches and glory of His Majesty's crowns seem chiefly to depend."[1] "If it prove or be thought," wrote the Duke of Ormond, "that Ireland's being above water hurts England, some invention must be found to sink it."[2]

In 1660 the new English tariff was drawn up, and high duties were placed on woollen goods imported from abroad. These duties were so high as to amount practically to a prohibition, and thenceforth Irish woollen goods were not able to find a market in England.[3]

Three years later the attack on Ireland's trade with the English Plantations began. The original Navigation Act of 1660[4] did not draw any distinction between English and Irish shipping, and was consequently welcomed in the Irish Parliament.[5] But in 1663 was passed the new Navigation Act,[6] which provided that no commodities, except horses, victuals, servants, and salt for the Newfoundland and New England fisheries, should be exported direct from Ireland to the English Plantations. In 1670 a further Act,[7] was passed prohibiting the direct exportation from the English Plantations to Ireland of sugar, tobacco, cotton, wool, indigo, and fustick or other dyeing wood.

As a matter of fact, these restraints were of very little importance to Ireland at the time, as the principal article exported to the Plantations was provisions, which were excluded from the Acts. It was not until the eighteenth century that the full effects of the Navigation Acts came to be felt. So long as Ireland's foreign trade was not interfered with, the only result of hampering Irish trade with the Colonies was to strengthen the bonds which

---
[1] Temple, *Essay on Trade of Ireland*, 1673.   [2] *Ormond MSS.*, vol. v., p. 417.
[3] Murray, *Commercial Relations*, p. 47.   [4] 12 Car. II., c. 18.
[5] *I.C.J.*, vol. i, pp. 420, 574-5.   [6] 15 Car. II., c. 7.   [7] 22 & 23 Ch. II., c. 26.

united Ireland with France and Spain. Nevertheless, the Acts were to some extent detrimental to Irish trading interest even at the date at which they were passed, and were much complained of. "The late Act," wrote Sir George Rawdon, "requiring all things to be landed in England which come from the Plantations will undo this Kingdom";[1] and again, "We are more undone in this poor Kingdom by the late Act than by the Cattle Act. . . . . We want a Lauderdale, etc., at court for the watching for Ireland as they do for Scotland."[2] "These laws laid Ireland under restraints highly prejudicial to her commerce and navigation. From those countries the materials for shipbuilding and some of those used in perfecting their staple manufactures were had; Ireland was by those laws excluded from almost all the trade of three-quarters of the globe, and from all direct beneficial intercourse with her fellow subjects in those countries which were partly stocked from her own loins. But still, though deprived at that time of the benefit of those colonies, she was not then considered as a colony herself, her manufacturers were not in any other manner discouraged, her ports were left open, and she was at liberty to look for a market among strangers, though not among her fellow subjects in Asia, Africa or America."[3]

Much more important than the Navigation Acts were the Acts which prohibited Ireland from sending her cattle into England. Before these Acts, three-fourths of the trade of Ireland was with England,[4] and, therefore, the blow inflicted by these Acts was aimed at the destruction of the major part of Irish trade; but, as we shall see, the ultimate effect of the Acts was to change the nature and the direction of the Irish export trade, to its great benefit. In June, 1663, a bill was introduced into the English House of Commons to restrain the importation of fat cattle and sheep from Ireland and Scotland. It

[1] C.S.P., Dom., 1671-2, p. 42. [2] C.S.P., Dom., 1671, p. 585.
[3] Hely Hutchinson, *Commercial Restraints*, p. 123.
[4] C.S.P., Ire., 1662-6, p. 693.

had been found, on inquiry, that each year about sixty-one thousand head of cattle were imported from Ireland, and it was thought that any further reduction of the price of English cattle alleged to be caused by this large importation would be prevented by the prohibition of Irish cattle from entering England after the first of August in each year. This bill passed without any serious opposition.[1]

This Act produced very serious results in Ireland, and formed the subject of a protest on the part of the Council of Trade, who represented that the restraint on exportation occasioned grievous suffering from one end of the Kingdom to the other; that many tenants were unable to pay their rents and were throwing up their leases; that the revenue would suffer from the diminished customs; that the restraints had put a great stop to trade, live cattle being the chief staple commodity of Ireland; and that it was impossible to substitute a trade in dead cattle and provisions owing to the backwardness of the country and the shortage of shipping. It was also represented that the restraint would in the end prove very injurious not only to Ireland but to England as well.[2] This representation was corroborated by the Irish Privy Council, who stated that all parts of Ireland were plunged into a sad degree of poverty; that vast quantities of cattle lay on their owners' hands for want of buyers; and that the farmers could not continue their tillage for want of money to pay wages.[3] These representations were laid before the King by the Duke of Ormond, who pointed out that Ireland had not yet recovered from the devastation of the Rebellion; that the Irish had no manufactures, and that of their vendible commodities cattle were the most lucrative; and that they could not start a provision trade with France or Holland owing to the war. Ormond reminded the King of the many restraints already in existence on the export of wool and linen yarn. He also pointed out that

[1] Carte, vol. iv, p. 234; 15 Car. II., c. 8.   [2] Carte, vol. iv, pp. 235-6.
[3] Carte, vol. iv, pp. 236-7.

the alleged reason for the new restraints, namely, the fall in the rents of land in England, was based on a misconception; that the fall of rents had been much greater than the value of the cattle imported, and must therefore have been brought about by other causes as well, such as "the revival of Lent after long disuse; the excessive drought of the last summer which hindered cattle from grazing; the retreat of abundance of sectaries to the American Plantations; the plague, which had destroyed a multitude of mouths that used to consume provisions; the obstruction given to trade by the war with Holland; the laying aside of the old hospitality used in the country; and the general decay of English manufactures."[1] As a matter of fact, it was afterwards admitted, even by the advocates of English interests, that the importation of Irish cattle had had nothing to do with the lowering of rents in England:—"The value of the oxen and sheep ever imported in one year out of Ireland into England was never worth more than £80,000, nor above the one-hundredth part of the rents of England, nor above the one-hundredth part of the butcher's meat consumed."[2]

The King was quite convinced that the restraint was unwise both from the English and Irish standpoint, but did not dare to dispense with the Act, as the matter was looked on as one of importance by the English House of Commons.[3] Ireland was thus plunged into distress. "The poverty," wrote Lord Anglesea, "caused by the prohibition against the export of cattle is so universal that it is past the skill or power of the Government to supply a remedy";[4] and Lord Orrery wrote in the same year that nothing but ruin stared Irish landlords and tenants in the face.[5] "Rents did not rise in England since the restraint; and yet Ireland was so much reduced thereby that the people had no money to pay the subsidies granted by Parliament, and their cattle was grown

---

[1] Carte, vol. iv, pp. 239-40.
[2] Collins, *A Plea for the Bringing in of Irish Cattle*, London, 1680.
[3] Carte, vol. iv, p. 242.   [4] C.S.P., Ire., 1666-9, p. 131.
[5] C.S.P., Ire., 1666-9, p. 166.

such a drug, that horses which used to be sold for thirty shillings were now sold for dogs-meat at twelvepence apiece, and beeves that before brought fifty shillings were now sold for ten shillings."[1]

But the English Parliament was not yet satisfied, and when it reassembled in October, 1665, a bill was brought in, not only to restrain the import of Irish fat cattle for certain parts of the year, but to prohibit that of all cattle from Ireland, fat and lean, dead and alive, at all times, and also sheep, swine, beef, pork and bacon. Owing to the vigorous opposition which the bill met with in the House of Lords, it was postponed until the next session.[2]

In the following year, however, the bill was reintroduced, and, in spite of a vigorous opposition in the House of Lords, was passed.[3] Needless to say, the passage of the bill met with the greatest resistance from all who were animated by the desire to consult the interests of Ireland. " Our next and indeed our greatest apprehensions," wrote Ormond, " are from the meeting of your Parliament, when it is feared the bill for the total prohibition of the export of cattle from Ireland to England will be proposed. This Bill is ruinous to all classes here."[4] " It will be some evidence," Ormond again wrote a month later, " of the ruin that must fall upon Ireland if the Act pass, that I shall ask that a proviso may be inserted in it giving the Lord Lieutenant power to license the exportation of twenty thousand, ten thousand, or five thousand head of cattle for the King's particular provision, my aim being by this means to make money of part of what my tenants can give me for rent."[5] " We shall shortly know our doom, as it may properly be called, concerning the prohibition of our trade with England . . . . what destruction the prohibition will bring upon Ireland, I think we have made evident to the King. He alone can judge whether the infallible destruction of a whole Kingdom and people

[1] Carte, vol. iv, p. 258. [2] Carte, vol. iv, pp. 242-5. [3] Carte, vol. iv, pp. 242-69.
[4] C.S.P., Ire., 1666-9, p. 76. [5] C.S.P., Ire., 1666-9, p. 209.

is to be admitted to prevent the possible damage which may fall in some proportion upon some persons in another."¹ "You cannot imagine," wrote Lord Dungannon, "how people's hearts are raised at the stop which we now understand is put upon the cow bill, wishing for good out of it; however, we prepare for the worst."² The Irish case against the Cattle Act is given very fully and ably in a letter from the Lord Lieutenant and Irish Privy Council to the King, which is too long to quote here in full, but is printed in Appendix IV.

The fact that the bill was calculated to injure English as well as Irish interests, did not make the slightest impression on those who were intent on passing it. As in the case of some more recent measures of the English Parliament, members were quite willing that England should suffer if such were the price which had to be paid for the gratification of inflicting a wrong upon Ireland. "No reasoning or arguments were heeded, or attempted to be answered; and, when Lord Anglesea and others laboured in private discourses to rectify the mistakes of such as were most furious in that affair, it appeared clearly enough that the Bill was carried on more out of wantonness and a resolution taken to domineer over that distressed Kingdom of Ireland, than any real belief that it would raise their rents."³ The attitude of the House of Commons towards Ireland may be judged by the fact that it actually passed an amendment allowing cattle to the imported from Scotland.⁴ This Act was further extended by later Acts which prohibited the import of Irish mutton, lamb, butter, and cheese.⁵ In 1668 the import of Irish cattle and horses into Scotland was prohibited.⁶

The immediate consequences of the Cattle Act were extremely injurious to Ireland, and numerous attempts

¹ C.S.P., Ire., 1666-9, p. 217. ² C.S.P., Ire., 1666-9, p. 268. ³ Carte, vol. iv, p. 263.
⁴ Carte, vol. iv, pp. 263-4. A full account of some interesting speeches delivered in the English Parliament against the Act is printed in C.S.P., Ire., 1666-9, pp. 533-42.
⁵ 18 Car. II, c. 23; 20 Car. II, c. 7; 22 Car. II, c. 2; 22 & 23 Car. II, c. 2.
⁶ *Catalogue of Tudor and Stuart Proclamations*, Bibliotheca Lindesiana, pp. 370-2.

were made to evade the provisions of the Act by the clandestine landing of cattle in England.[1] It is interesting to note that Ormond did not feel himself bound to discountenance these attempted evasions, on the ground that an Act of the English Parliament could impose no duty on the Irish executive.[2] In 1671 sheep from Ireland were still being illegally landed in South Wales;[3] and in 1675 the landing of Irish cattle at Chester was generally connived at.[4] It was probably the knowledge that a good many Irish cattle were finding their way into England that caused the yokels to attack and destroy any cattle which they saw going along the road which they suspected of being tainted with an Irish origin, as happened at Brough in 1677.[5] It is hard to believe that these rustics were actuated by any abstract principles of political economy, even mistaken ones; and their hostility to these poor beasts was probably a symptom of the widespread hatred of all things Irish that prevailed in England at that time. Their reception of Irish cattle was very similar to their reception of the Irish soldiers who were brought to England by James II. The great landowners, however, did not always study the feelings of their inferiors, and obtained licences to transport thousands of head of the despised Irish breed, when it suited their convenience or their pocket to do so.[6]

The Irish did their best to make up for the great loss inflicted on them by the Cattle Act. "To procure a rent for their cattle some proposals were made to the Lord Lieutenant for carrying them alive to Rotterdam; but the length of the voyage, the uncertainty of winds, and other expenses considered, it was found that they could not be delivered there as cheap as the Dutch could be supplied with them from Holstein.[7] No

---

[1] C.S.P., Ire., 1666-9, pp. 303, 305, 308, 311. Three contemporary accounts of the immediate effects of the Act in Ireland are printed in Appendix v.
[2] C.S.P., Ire., 1666-9, p. 338.   [3] C.S.P., Dom., 1671-2, p. 495.
[4] C.S.P., Dom., 1675-6, p. 297; C.S.P., Dom., 1679-80, p. 20.
[5] C.S.P., Dom., 1677-8, p. 414.   [6] C.S.P., Dom., 1675-6, p. 116; C.S.P., Dom., 1678, p. 486.
[7] A certain number of live cattle were exported abroad, C.S.P., Dom., 1675-6, p. 397.

way appearing of making any advantage of their cattle until after they were dead, and the freight to foreign parts amounting in general to two-thirds of the value of the commodities, other proposals were made for the increase of shipping."[1] The Duke of Ormond also attempted to encourage the woollen and linen industries,[2] and the people entered into non-importation agreements.[3]

The King meanwhile did much to relieve the condition of affairs in Ireland by directing that all restraints upon the exportation of commodities of the growth or manufacture of Ireland to foreign countries, either at peace or war with England, should be taken off, and by giving leave to the Irish Privy Council to prohibit the importation of woollen and linen manufactures and other commodities out of Scotland, which had copied England in prohibiting the import of cattle from Ireland.[4] This relaxation proved of immediate benefit: "All our seaports are full of trade, great store of our country commodities being daily shipped off."[5]

Although the immediate effects of the Cattle Acts were so detrimental to Irish interests, yet the ultimate results were beneficial to Ireland and detrimental to England. "The Act," says Carte, "has since proved of advantage to Ireland, but very destructive to the trade of England."[6]

"It was to this party of men that their country owes the act of prohibiting Irish cattle—an act which, how grievous soever it was to Ireland for a time, and in that particular juncture, hath since proved of advantage to that kingdom, but very destructive to the trade of England. The Irish till then had no commerce with this kingdom, and scarce entertained a thought of trafficking with other countries. They supplied us with their native commodities, which made work cheap, and carried off our artificial ones to a value which exceeded that of their

[1] Carte, vol. iv, p. 282.   [2] Carte, vol. iv, pp. 283-5.
[3] C.S.P., Ire., 1666-9, p. 293.   [4] Carte, vol. iv, p. 288.
[5] C.S.P., Ire., 1666-9, p. 479.   [6] Vol. iv, pp. 273-4.

own; so that they were rather impoverished than improved by the traffick. The English were undoubtedly the gainers by this mutual trade, from which they now so wantonly cut themselves off by forbidding the principal part of it, and rendering the rest impracticable. They soon felt the consequences of this unhappy step; the Irish, forced by their necessities to be industrious, set themselves to improve their own manufactures, and carried their trade to foreign parts, from whence they brought those commodities which they used to take from England. In this country the price of meat rose considerably as soon as the Act passed; even before the end of this session of parliament (which broke up on Feb. 9) the price of labour and rates of wages were thereby enhanced; and the wool of Ireland, which never before had any vent but in England, being now carried abroad, foreigners were thereby enabled to set up woollen manufactures, and, by the cheapness of labour in their country, undersell us in that most beneficial branch of our commerce. The English have since sufficiently felt the mischiefs of this proceeding, which were, in truth, obvious enough to be foreseen at that time by a man of common understanding, but it will puzzle the wisest to find a remedy to remove them now they have actually happened. It would be well if any experience could make them wiser, and dispose them to treat Ireland better."[1]

The principal benefit which Ireland derived from the Cattle Act was that it precipitated the development of her foreign provision trade. The fact that, instead of exporting live cattle to England, Ireland was beginning to export dead cattle, and the finished products thereof, to foreign countries, soon became apparent. For instance, in 1668, a year after the passing of the Act, a ship left for France laden with large quantities of butter, tallow, hides, and bones.[2] The following table of exports illustrates this change:—[3]

[1] Carte, vol. iv, pp. 273-4. [2] C.S.P., Ire., 1666-9, p. 615.
[3] C.S.P., Ire., 1669-70, p. 54.

	Year ending 25th December, 1665.	Year ending 25th December, 1669.
Beeves, no.	37,544	1,054
Beef, barrels	29,204	51,793
Butter, cwt.	26,413	58,041
Cheese, cwt.	318	1,227
Tallow, cwt.	21,003	38,183
Candles, cwt.	1,330	3,473
Hides, no.	106,344	217,046
Calf Skins, doz.	612	1,731

Sir William Temple was much impressed in 1673 by the growing trade in Irish provisions, of which he wrote as follows :—

"Until the transportation of cattle into England was forbidden by the late Act of Parliament, the quickest trade of ready money here was driven by the sale of young bullocks, which for four or five summer months of the year were carried over in very great numbers, and this made all the breeders in the kingdom turn their lands and stock chiefly to that sort of cattle. Few cows were bred up for the dairy, more than served the consumption within; and few oxen for draught which was all performed by rascally small horses; so as the cattle generally sold either for slaughter within or exportation abroad, were of two, three or at best four years old, and those such as had never been either handled or wintered at hand meat, but bred wholly upon the mountains in summer, and upon the withered long grass and the lower lands in the winter. The effect thereof was very pernicious to this kingdom in what concerned all these commodities; the hides were small, thin and lank; the tallow much less in quantity and of quicker consumption. Little butter was exported abroad, and that discredited by the housewifery of the Irish, in making it up; most of what was sent going from their hands, who alone keep up the trade of dairies, because the breed of their cattle was not fit for the English markets. But, above all, the trade of beef for foreign exportation was prejudiced and almost sunk; for the flesh being young, and only coarse fed (and that on a sudden by the sweetness of the summer pasture, after the cattle being almost starved in the winter) was then light and moist, and not of a substance to endure the salt, or be preserved by it, for long voyages, or a slow consumption. Besides, either the unskilfulness, or carelessness, or knavery of the traders, added much to the under value and discredit of these commodities abroad; for the hides were often made up very dirty, which increased the weight, by which that commodity is sold when it goes in quantities abroad. The butter would be better on the top and bottom of the barrel, than in the middle, which would be sometimes filled up, or mingled with tallow; nay, sometimes with stones. The beef would be so ill-chosen, or so ill-cured, as to stink many times before it came so far as Holland, or at least not prove a commodity that would baffle the first charge of the merchant before it was shipped. Nay, I have known merchants there having to throw away great quantities after having lain long in their hands without any market at all.

"After the Act in England had wholly stopped the transportation of cattle, the trade of this kingdom was forced to find out a new channel; a great deal of land was turned to sheep, because wool gave ready money for the English markets, and by stealth for those abroad. The breeders of English cattle turned much to dairy, or else by keeping their cattle to six or seven years old, and wintering them dry, made them fit for the beef-trade abroad; and some of the merchants fell into care and exactness in barrelling them up; and hereby the improvement of this trade was grown so sensible in the course of a few years, that in the year 1699 some merchants in Holland assured me, that they had received parcels of beef out of Ireland which sold current, and very near the English; and of butter which sold beyond it; and that they had observed it spent as if it came from the richer soil of the two. It is most evident that if the

Dutch war had not broken out so soon after the improvements of all these trades (forced at first by necessity, and growing afterwards habitual by use), a few years would have very much advanced the trade and riches of this kingdom, and made it a great gainer, instead of losing by the Act against transportation of their cattle. But the war gave a sudden damp to this and all other trade, which is sunk to nothing by a continuance of it."[1]

In 1680 it was said that the Irish undersold the English with provisions at Dunkirk.[2] " There are seldom less than twenty Irish ships at Dunkirk, laden with beef, tallow, hides, and leather, much butter and some wool .... divers other Irish ships furnish Ostend; and many more are seen at Nantes, Burnes, and Rochelle." " Last Friday, coming to the pier of Penzance, was cast away an Irish vessel of about sixty tons, laden with tallow, beef, and hides for Ostend."[3] The English butter trade in France, Flanders, Spain and Portugal was also ruined.[4] It was not alone in foreign countries that the Irish provisions undersold and displaced the English; the same thing happened in the English and foreign plantations.[5] " The Islands and Plantations in America are in a measure wholly sustained by the vast quantities of beef, pork, butter, and other provisions of the product of Ireland."[6] English provision merchants in the Canaries were ruined by the competition of the Irish.[7]

One very valuable species of trade which was to some extent lost to England by the increased Irish output of provisions was the victualling of English and foreign ships. " The East and South parts of England have lost the victualling of the Dutch and of our own merchantmen; and our ships for the most part Westerly and Southerly bound victual here but for six weeks, and take in the

[1] Temple, *Essay on the Trade of Ireland*, 1673.
[2] *A Plea for the Bringing in of Irish Cattle*, by Collins, London, 1680.
[3] C.S.P., Dom., 1675-6, p. 6.
[4] Coke, *Equal Danger of Church, State and Trade of England*, London, 1675. The Irish export of butter attained very large dimensions, and was put under regulations by a Statute of 1698, 10 Will. III, c. 2. Proposals by Sir William Temple for the advancement and regulation of the growing provision trade are printed in Appendix vi.
[5] Murray, *Commercial Relations*, p. 35.
[6] *The Interest of England in the Preservation of Ireland*, by G. P., London, 1689.
[7] C.S.P., Dom., 1673-5, p. 444.

rest of their provisions in Ireland or Irish provisions in Spain."¹ The volume of Irish-owned shipping was also increased.² Ireland, moreover, was placed in a position favourable in time of war. " It is not the King of France who supplies the Irish," we read in 1691, " he being at not one penny's expense to do it, but it is the advantageous trade hither for hides and tallow that does it, and which the merchants can make such vast advantage with so little hazard, they will furnish them to the end of the world."³

Of course, this diversion of Irish trade from England to foreign countries had another result, namely, the practical cessation of the importation of English goods into Ireland. Naturally, the ships employed in exporting Irish provisions did not return empty; and thus it came about that Ireland began to buy in Spain and Holland the commodities which she had theretofore purchased in England. Before 1663 Ireland had on an average imported about £200,000 worth of English goods annually ; but in 1675 the value of the goods so imported was not more than £20,000.⁴ This also threw out of employment many English ships and seamen who had been engaged in the trade across the Irish channel.⁵

Another result of the Cattle Acts was that much land which had formerly been devoted to the grazing of cattle was given over to sheep. This had two results—first the growth of the Irish woollen manufacture, and secondly the vastly increased smuggling abroad of Irish wool which enabled foreign woollen manufacturers to undersell and successfully compete with the English. As early as 1669 it was remarked that the exportation of Irish wool to France, and the consequent injury to the English woollen industry had much increased since the Cattle Act.⁶ This formed a matter of constant complaint

---

¹ Collins, *A Plea for the Bringing in of Irish Cattle*, London, 1680.
² *Remarks on the Affairs and Trade of England and Ireland*, London, 1691.
³ C.S.P., Dom., 1690-1, p. 265.
⁴ Coke, *England's Improvement by Foreign Trade*, London, 1675.
⁵ Murray, *Commercial Relations*, pp. 36-7 ; *Remarks on Affairs and Trade of England and Ireland*, London, 1691.
⁶ *England's Interest Asserted in its Native Commodities, and More Especially of the Manufacture of Wool*, London, 1669.

in England, and there was much agitation for a repeal of the obnoxious Acts, which were alleged to be causing the complete ruin of the woollen trade, as well as raising the price of beef in England.[1] "Irish cattle being prohibited they breed more sheep and bring in more wool into England besides what they send beyond sea; which will infallibly bring our lands in England as low as those in Ireland, *i.e.*, to as low a rent and to as few a years value in the purchase, nay lower, if they be suffered first to glut England with their wool, and then to furnish the markets beyond sea; yet we prohibited the same privilege which is our present condition, and undoubtedly the forbidding Irish cattle has been of vast inconvenience, not only to the best of England, the feeding lands, but to all of it in general. By lessening the value of our wool, in which even the breeding lands receive more loss by the low price of their wool than they reap advantage by this Act in the price of their cattle, this Act is also injurious to the nation by sending our own foreign ships to victual in Ireland; by the want of returns from thence; by loss of our trade for hops, hides, butter, cheese, etc., which trades are now taken up by the Irish to the ruin of many counties of England; by discouraging navigation, for it is said that a hundred of our ships were continually employed in this traffic of lean cattle. And lastly by discouraging our clothiers and other manufacturers, who since they must live out of their labours, the dearer they pay for their diet the more they must have for their work. The Irish Act, therefore, making our beef dear, yet the Dutch having it from Ireland delivered in Holland at about a penny a pound, they may afford their cloth cheaper than possibly we can; which will speedily enable them to get from us also our foreign clothing trade, and be an irreparable damage to this Kingdom, if the Parliament in their wisdom do not prevent it. Thus the Act which in its preface designs the advancing our rents and enriching England has

[1] C.S.P., Dom., 1673-5, p. 183.

lessened and impoverished both; has compelled Ireland to seek a way to live without us; has made it almost independent of England; has in fine almost ruined both nations.''[1]

" I know no law," wrote Cary in 1695, " in my time that hath been more pernicious to the traffic of England; 'twas this first put those of Ireland on that trade which hath since almost eat out ours; 'twas this set them on manufactures which were so far advanced before the late troubles that the sales of our market came to a thousand pounds per week; 'twas this that hath produced such great quantities of wool in Ireland as have at least equalled, if not exceeded England, for the greatest part of the lands of that Kingdom, by reason of the thinness of its inhabitants, being turned rather to pasture than tillage, and that prohibition discouraging the raising of black cattle, put the people on stocking them with sheep.''[2]

Thus England, and not Ireland, was the ultimate loser by the Cattle Acts, a result which, in after years, was freely admitted and deplored by English writers.[3] " The owners of breeding lands have since the prohibition not gotten above 10s. per head more for their cattle than before it, which the owners of the feeding lands have paid them and lost. Moreover, the mariners of England have lost the getting of 9s. 6d. per head for freight and primage, and the people of England have lost 4s. 6d. a head more for driving and grazing; the King hath lost 3s. 6d. per head for customs on both sides, besides officers fees; and the traders in hides and tallow have lost what they might have gained out of 15s. per head; and the merchants and artisans of England have lost yearly what they might have gained by £140,000 worth of English manufactures; the wool growers of England have lost as much as their wool is fallen by reason of the extraordinary sheepwalks now in Ireland;

[1] *Reasons for a Limited Exportation of Wool*, London, 1677.
[2] Cary, *Essay on the State of England in Relation to its Trade*, Bristol, 1695.
[3] Sir William Temple's expression of this view is printed in Appendix vii.

the landlords of Ireland resident in England have lost five per cent. extraordinary for exchange of money. Lastly, the bulk of the people of England have lost one half-penny for every pound of fresh meat they have spent, amounting for all England into more than two millions per annum, of which great sum the owners even of breeding lands have paid three times more in the enhancement of wages and manufactures than they got by the raised price of their cattle above-mentioned."[1]

" The unfortunate Act (for so I must humbly crave leave to call it) was made against the importing the Irish cattle, upon supposition that it would raise the price of land here in England, whereas the quite contrary effect has been too much experienced (viz.:) that it hath laid such a foundation for the impoverishing of England as will not quickly be recovered."[2]

The Dutch War must have had a very serious effect on the trade of Ireland while it lasted. " This war has had a more particular and mortal influence upon the trade of this country than upon any other of his majesty's Kingdoms. For by the Act against transfer of cattle into England the trade of this country which was wholly thither before was turned very much into foreign parts; but by this war the last is stopped, and the other not being open, there is in a manner no vent for any commodity but of wool. This necessity has forced the Kingdom to go on still with their foreign trade; but that has been with such mighty losses by the great number of Dutch privateers plying about the coasts, and the want of English frigates to secure them that the stock of the Kingdom must be extremely diminished."[3] " I wish we had a more probable prospect of the end of the war," wrote Ormond in the same year. " Not only the people of England, I believe, feel the ill-effects of it in their trade, but those in this country bear a great share, and

[1] Collins, *A Plea for the Bringing in of Irish Cattle*, London, 1680; *England's Interest by the Improvement of the Manufacture of Wool*. By W. C., London, 1669.
[2] *A Summary of Certain Papers about Wool*. By W. C., London, 1685. A long and interesting memorandum advocating the repeal of the Cattle Acts in the interest of England is printed in Appendix viii. [3] Temple, *Essay on Trade of Ireland*, 1673.

for the present it puts a stop even within the Kingdom to many improvements of which I am confident that no country in Europe is more capable than this."¹ Theoretically, the Irish were permitted to trade with the countries at war with England as a consolation for their losses from the Cattle Acts; but in practice this permission was of no value, as the Irish coasts were harassed by Dutch privateers, who drew no distinction between the different classes of the King's subjects.² At a later date the war between France and Spain temporarily paralysed Irish trade, as no protection was afforded to it by the English fleet;³ and in 1679 the rumours of trouble in Scotland seriously affected Irish trade.⁴

In 1682 there appeared a book entitled, "The Interest of Ireland in its Trade and Wealth Stated," and from this we learn that, although Irish trade was continuing to grow, it was hampered in several ways. One cause of this was the high rate of interest on money in Ireland, which could not be procured at less than ten per cent., whereas in Holland it might be had at three. This kept a great many people from setting up as traders. The low price of land operated in the same direction, nobody being willing to embark his money on trade risks, when he could purchase land at ten years' purchase, and thus obtain a high return on his money in the shape of ground rents.⁵ Trade was further hampered by " the low esteem the generous calling of a merchant hath in this country where every fiddling shopkeeper and pettifogger is called a merchant; Ireland breeds merchants as beggars do lice from its poverty and idleness." The credit of the Irish merchant abroad was unfortunately not high, but this was caused not by any dishonesty on his part, but by the difficulty which he experienced in getting paid what he was owed by the gentry.

---
[1] C.S.P., Dom., 1673, p. 585.   [2] C.S.P., Dom., 1673, pp. 180, 189-90; C.S.P., Dom., 1677-8, p. 56.
[3] C.S.P., Dom., 1673-5, p. 21.   [4] C.S.P., Dom., 1679-80, pp. 173-9.
[5] This is also noticed in a pamphlet entitled, *The Character of the Protestants of Ireland*, London, 1689.

One result of the high rate of interest on money was that the Irish merchants tended to sink to the position of being merely factors of English and foreign merchants, who could raise their capital at an easy rate.[1] "The petty shopmen, traders in small towns, and country dealers are only hawkers, precursors, and brokers for the greater merchants in the cities and big towns; and they ordinarily make themselves factors for the merchants in England."[2]

As soon as Tyrconnell became Lord Lieutenant, he turned his attention to the improvement of the trade of the country; this was probably one of the reasons for the intense unpopularity with which his vice-royalty was regarded in England. "Ireland is in a better way of thriving under the influence of a native governor than under any stranger to us and our country. A man altogether of English interest never did and never will club with us or project anything for us which will tend to our advantage that may be the least bar or prejudice to the trade of England, which is the only nation in the world that injures our trade. The embargo upon our West India trade is without parallel, considering we are the same prince's subject; the like upon our Irish cattle no less; an Act made by the interest and faction of a few landlords, not only to the prejudice of a Kingdom as big and fertile almost as England, but to the disaccommodation of many thousands in England as often as they are hungry."[3]

As we have seen above, Irish shipping was completely ruined during the Rebellion. In 1665 it was stated that Ireland had no ships of her own;[4] "Ireland since their decay by the late horrid rebellion having almost no shipping left of her own."[5] In 1663 there were twenty-nine ships owned by inhabitants of Belfast; one of 200 tons, one of 150 tons, one of 120 tons,

---

[1] Lawrence, *op. cit.*, vol. ii, p. 13; Temple, *Essay on Trade of Ireland*, 1673.
[2] *The Interest of England in the Preservation of Ireland.* By G. P., Esq., London, 1689.
[3] *A Vindication of the Present Government of Ireland in a Letter to a Friend*, London, 1684. [4] Carte, vol. iv, p. 235. [5] C.S.P., Ire., 1666-9, p. 187.

and the rest very small, totalling 1,102 tons; twenty years later the tonnage of this port had increased to 1,527 tons.¹ In 1668 a computation was delivered to the Council of Trade, showing that the exportation from Ireland amounted to about £450,000 per annum, and there was paid to English ships for the freight of Irish imports and exports £130,000 per annum.² Sir William Petty was of opinion that the Irish had no inducement to build ships, as they could hire them very cheap from the Dutch. " Some have thought that little shipping belongs to Ireland by the great policy of the English who would keep the chain or drawbridge between the two Kingdoms on the English side; but I never perceived any impediment of building, or having ships in Ireland, but men's own indisposition thereunto, either for not having stock for so chargeable a work, or not having workmen enough of sorts to fit out a ship in all particulars; as for that, they could hire ships cheaper from the Dutch than to build them. Nevertheless at this day there belongs to several ports of Ireland vessels between 10 and 200 tons, about 8,000 tons of several sorts and sizes."³ "I doubt," wrote Sir William Temple, " there is hardly any other country lying upon the sea coast, and not wholly out of the way of trade, which has so little shipping of its own as Ireland, and which might be capable of employing more. The reason of this must be in part the scarcity of timber proper for this built, but more the want of merchants and uncertainty of trade in the country."⁴

The increase of foreign trade, which, as we have seen, was the direct result of the Cattle Acts, produced a slight increase in the volume of Irish shipping.⁵ The Dutch War also helped by compelling Irish merchants to look for Irish carriers.⁶ The increase, however, must have been inconsiderable; in 1682 Lawrence complained that " the loss by trading in foreign bottoms is a vast charge

[1] Benn, *History of Belfast*, p. 310.
[2] *Remarks on the Affairs and Trade of England and Ireland*, London, 1691.
[3] *Political Anatomy*, p. 110. [4] *Essay on the Trade of Ireland*, 1673.
[5] *Remarks on Affairs and Trade of England and Ireland*, London, 1691.
[6] C.S.P., Dom., 1675-6, p. 242.

to this Kingdom, computed to amount to at least £60,000 per annum, besides the loss of the seamen's habitations and families expenses, and of many artists engaged in rigging and repairing of ships."[1] The shipping that belonged to the port of Dublin in 1687 was "not worth speaking of";[2] and in 1691 Dublin did not own more than ten ships.[3] It was said in 1698 that Dublin had not a single ship, while Belfast and Cork owned but a few small craft, and there was not one large vessel in the whole Kingdom.[4] "Ireland hath such a provision of shipping," we read in 1697, "as may serve to begin a trade."[5]

It should be remembered that one of the measures passed by the much maligned Patriot Parliament was an Act for the re-development of Irish shipping. This Act was so wide in scope, and designed to be of such great benefit to the country, that the sections dealing with shipping are worth quoting in full:—

AN ACT FOR THE ADVANCE AND IMPROVEMENT OF TRADE, AND FOR ENCOURAGEMENT AND INCREASE OF SHIPPING, AND NAVIGATION.

(5) And for the more Encouragement of building good and serviceable ships, Be it Enacted by the Authority aforesaid, That any person or persons, who shall within the space of Ten Years, to commence the 24th of June, 1689, build or cause to be built within this Kingdom of Ireland, any Ship or Vessel above Twenty-five Tun, and under One Hundred Tun, Burthen, shall and may for the first three Voyages any such Ship or Vessel shall make out of this Kingdom, upon the said Ship or Vessel's Return from such Voyage back into this Kingdom, have, receive, or be allowed to his and their own proper Use one Eighth Part of the Duties of Customs and Excise, which shall be due or payable to the King, his

[1] Lawrence, *The Interest of Ireland in its Trade and Wealth Stated*, London, 1682, vol. i, pp. 13, 83.   [2] *A New Irish Prognostication*, London, 1689, p. 89.
[3] *Remarks on Affairs and Trade of England and Ireland*, London, 1691.
[4] *An Answer to a Letter from a Gentleman in the Country to a Member of the House of Commons relating to the Trade of Ireland*, Dublin, 1698. In 1685 about fifty ships, probably small ones, were said to belong to Cork, *Description of Ireland*, by Sir. R. Cox, Kilk. Arch. Jnl., 5th s., vol. xii, p. 353.
[5] *A Proposal and Considerations relating to an Office of Credit upon Land Security*, by Dr. Hugh Chamberlain, London, 1697.

Heirs, or Successors, for and out of all the Goods and Commodities so imported in such Ship or Vessel upon the said three first Returns, which such Ship or Vessel shall make into this Kingdom. And likewise, That any Person or Persons who shall within the said space of Ten years commencing, as aforesaid, build or cause to be built in this Kingdom any Ship or Vessel exceeding in Burthen One Hundred Tun, shall for the first four Voyages such Ship or Vessel shall make out of this Kingdom, and upon the said Ship or Vessel's Return from the said Voyages back to this Kingdom, have and receive to his and their own proper Use one Eighth Part of the Duties of Custom and Excise, which shall be due or payable to the King, his Heirs or Successors, for or out of the Goods and Commodities so imported into such Ship or Vessel upon the four first Returns such Ship or Vessel shall make out of this Kingdom.

(6) And to the end that Masters of Ships, Seamen, Mariners, Shipwrights, Carpenters, Rope-makers, and Block-makers may be encouraged and invited to come and dwell in this Kingdom, and that thereby Navigation may improve and increase, Be it further Enacted by the authority aforesaid That all and every Masters of Ships, and Shipwrights, Ship Carpenters, Seamen, Mariners, Rope-makers, and Block-makers, who are at present residing within this Kingdom, or who shall or do at any time from henceforth come and reside in this Kingdom of Ireland, and shall pursue and follow his Trade or Calling, shall and may from the time and space of Ten Years after his or their so coming into this Kingdom, be freed, exempted and discharged of, and from all sorts of Taxes, and Cesses, Watch, Ward and Quarterings of Soldiers and Officers in and throughout and Kingdom: And shall likewise have and be allowed his and their Freedom gratis in any Town, City, Seaport, Corporation or Borough, where he or they shall please to reside, and pursue his or their Calling and Trade.

(7) And be it further Enacted by the Authority aforesaid, That in the respective Cities and Towns of Dublin, Belfast, Waterford, Corke, Limerick and Gallway there shall be established, erected and settled, before the First day of December, 1689, in each of the said Towns and Cities, and so continued for ever hereafter, a Free School for Teaching and Instructing the Mathematicks, and the Art of Navigation; in every of such Schools there shall be placed and continued one or more able and sufficient Master or Masters for Teaching and Instructing the said Arts: And that every of the said Towns and Cities shall out of the Publick Revenue and Stock to them

belonging, or otherwise, settle and secure a reasonable Pension and Stipend for such Master or Masters, to be paid them Quarterly during his or their Continuance in such Employment or Employments.

By other sections of this statute, Irish trade was encouraged in still more important ways. The Navigation Acts were repealed, and a complete free trade was opened up between the Plantations and Ireland. The Act contained a provision that a large part of all duties paid on the importation of Plantation goods into Ireland should be remitted on re-exportation, and encouragement was thus given to the development of Ireland as the centre of a great ocean trade. If this wise and beneficial Act had been allowed to come into force by James's successor, Ireland would have been saved from much of the industrial depression which characterized the eighteenth century. It must be remembered that the same Parliament which passed this Act also established the principle of Irish legislative independence, and that, had not that principle been trampled on in the years following the Williamite victory, the worst measures of English commercial oppression could not have been enforced. Whatever may be one's opinion of the praise or blame that should be attached to the legislation of the Patriot Parliament, one thing at least is quite certain—that, had that legislation been allowed to take effect, Ireland, in the eighteenth century, would have been an incomparably richer and more prosperous country than it was.[1]

(10) *Industries in General.*

On the whole, it may be said that during the period from the Restoration to the Revolution, Ireland was not a manufacturing country, but was concerned rather with the production of food, and the raw materials of manufacture. This was natural, in view of the impoverished state that the country was in after the Civil Wars, which had almost depopulated it. The country, moreover, was

[1] The sections of the Statute of the Patriot Parliament dealing with trade not printed above are printed in Appendix ix.

poor, and the commodities consumed by the vast majority of the inhabitants were the simple necessaries of life. There was, therefore, no demand for the more highly-finished manufactured articles, and any industry which hoped to attain to a large scale in Ireland would have had to look abroad for its principal markets. The English market, however, was closed by the high duties imposed at the Restoration; the Colonies were practically closed by reason of the Navigation Acts; and the infant industries of Ireland would have found it extremely difficult to compete with their richer English rivals in the markets of foreign countries. Another drawback in Ireland was the difficulty of obtaining money at a reasonable rate, a difficulty which the more thoughtful students of affairs thought could be removed by the creation of a land bank.[1]

The Government always professed anxiety to promote the growth of Irish industry, however much its professed intentions were belied by its actions. The Commonwealth Government gave orders to the Irish Commissioners to encourage the manufacture of woollen and linen cloth,[2] and appointed a committee to consider how manufactures might be encouraged. Again, Charles II. repeatedly ordered his Deputies to exert themselves to increase the industries of Ireland, provided always that they did not conflict with English interests. The encouragement given to Irish manufactures by the Duke of Ormond is well known.

One settled policy of the Government in this respect was the encouragement given to foreigners to settle in Ireland. In 1661 the King ordered Ormond to see that this was done,[4] and in the following year an Act was passed,[5] which provided that all Protestant traders, merchants and artisans who for seven years should transport their stocks and families to Ireland should be granted free naturalization. This policy was consistently carried

---

[1] Lawrence, *The Interest of Ireland in its Trade and Wealth Stated*, London, 1682.
[2] Dunlop, p. 460.   [3] *Ibidem*, p. 578.   [4] C.S.P., Ire., 1660-2, p. 410.
[5] 14 & 15 Car. II., c. 13.

out by Charles II., both in the interests of the Irish revenue and of the Protestant religion, and was also advocated by James II.[1]

Of course, the growth of industries during this period was hampered to a large extent by the fact that Ireland was hindered from competing in the English or colonial markets. To develop an export trade to foreign countries was a difficult matter under these conditions, as Irish goods were everywhere met by the products of older established and larger industries, and Ireland was not in a position to produce on a large scale. It might have been possible to overcome these difficulties if the Irish producer had been secured in the certainty of a market in Ireland for the greater part of his produce, but it must be remembered that the greater part of the Irish population was self-supporting and bought a very small number of commodities with money. " If it be true," wrote Petty, " that there are but 16,000 families in Ireland that have above one chimney in their house, and above 180,000 others, it will be easily understood what the trade of the latter sort can be, who use few commodities; and those such as almost everyone can make and produce. That is to say, men live in such cottages as themselves can make in three or four days; eat such food as they buy not from others; wear such clothes as the wool of their own sheep, spun into yarn by themselves, doth make. . . . The whole annual expense of such a family seems to be but about 52 shillings per annum."[2] " A pot, a griddle whereon to bake their bread, a little snuff, salt, and iron for their ploughs, is almost all they trouble the merchant or shopkeeper for."[3]

There is nothing to be said of the organization of industry in the seventeenth century in Ireland, except that no sign of the Industrial Revolution had begun to appear, and that the period was not marked by any noticeable advance in the methods of production. The following extracts

[1] *Laing MSS.*, vol. i, p. 447.   [2] *Political Anatomy*, p. 76.
[3] *Remarks on the Trade and Interest of England and Ireland*, London, 1691.

from works dealing with the economic development of England apply equally to Ireland, with the exception that there is no necessity to speak of the industrial groups in the more rural districts, for the reason that in the rural parts of Ireland, no industry was carried on, except the primitive supplying by each cottier family of its own immediate needs. " The century and a half which separated the death of Elizabeth from the accession of George III. is elusive of historical treatment. The surface of the national economy was strikingly little altered; new industries, indeed, arose, and new trades were opened up, but such progress was conducted for the most part on lines marked out already in the sixteenth century. There are no structural alterations of the larger kind. The relations, both social and economic, between employer and wage-earner, landlord, tenant and labourer, even at the close of the period were very much what they had been at its commencement."[1] " In the seventeenth century the advantages of the division of labour were widely recognized. In the towns there was a combination of the domestic with the manufacturing system, *i.e.*, certain operations were carried on by handicraftsmen grouped together in families or workshops while subsidiary processes were performed in the homes of the people. In the more rural districts the usual industrial group was the family. The dealer gave out raw material to be worked up by the craftsman and his wife and children. But there was no deep line of division between dealer and artisan or merchant and employed."[2]

(11) *The Woollen Industry.*

As we have seen, the woollen industry suffered equally with every other industry during the years of the Rebellion. In 1651 a proclamation was made prohibiting the further killing of sheep and lambs, as the breed was being destroyed by reason of the great destruction of live

---

[1] Meredith, *Economic History of England*, London, 1915, p. 181.
[2] Hewins, *English Trade and Finance*, London, 1892, p. 96.

stock during the war.¹ In time, no doubt, the great Irish sheepwalks became again covered with sheep, but the country was so depopulated and unsettled that the wool was not worked up into cloth. In 1659, wool was said to be plentiful in the north, but was not worked up, and the establishment of a knitting industry was suggested.² In considering the woollen industry in the years 1660-90 we are, therefore, considering an industry which was practically starting from nothing.

The manufacture was interfered with by two sets of restraints, those on the export of wool, and those on the export of cloth. Strictly speaking, wool could be legally exported from Ireland to any part of the world with a license, but in practice it could only be transported to England, as licenses to transport elsewhere were extremely difficult to obtain, whereas licenses to transport to England were readily granted. The State Papers contain numerous references to licenses to transport to England; in July, 1661, one John Codd was authorized to send a hundred large bags,³ and later in the same year several other licenses were granted.⁴ A license, however, to export to Holland was refused, because the Dutch woollen manufacturers could not work up their Spanish wool unless it was mixed with Irish.⁵ The only example of a license to export a considerable amount of wool to foreign countries is that given in 1661 to Major Henry O'Neile, allowing him to send sixty thousand pounds of wool to France, Flanders, or any other place.⁶ This license was afterwards extended.⁷ In 1662 the King wrote to the Lord Lieutenant that he had made the discovery that if all the Irish wool were manufactured in Ireland, " the same would be of incomparable inconvenience to the trade of clothing in our Kingdom of England, considering that wools being cheaper in Ireland than here, our subjects there might undersell our

---

¹ Bagwell, vol. ii, pp. 245, 301. ² C.S.P., Ire., 1647-60, p. 694.
³ C.S.P., Ire., 1650-2, p. 309. ⁴ C.S.P., Ire., 1660-2, p. 378.
⁵ C.S.P., Ire., 1660-2, p. 385. ⁶ C.S.P., Ire., 1660-2, p. 305.
⁷ C.S.P., Ire., 1660-2, p. 230.

subjects here, and consequently much decay the trade of clothing in this our Kingdom, which we must not admit "; and he, therefore, authorized the Lord Lieutenant to grant licenses for the export of wool to England, but not elsewhere;[1] and at a later date the Lord Lieutenant was encouraged to grant such licenses, and to pardon smugglers, lest the Irish wool should be worked up in Ireland.[2] A special office was established for registering all bonds regarding the export of wool, and for collecting such bonds when forfeited.[3] The importance attached to these restrictions may be judged from the fact that, when in 1667 the King authorized the Lord Lieutenant to make a proclamation taking off all restraints on the export of commodities from Ireland, the Duke of Ormond felt it his duty to except wool from that proclamation, as he could not believe that the King's English advisers would tolerate such a concession.[4] There was nothing peculiarly intolerant or oppressive in the English attitude in this matter, which was simply in accord with the notions of the time. Indeed, when the French in 1690 saw the possibility of becoming the masters in Ireland, they proposed to pursue a precisely similar policy; one of the suggestions of Avaux being that the export of Irish wool to France should be made free, but the export to England prohibited.[5]

The finished product of the woollen manufacture could in theory be exported to any part of the world, but in fact the English market was closed by the high duties which had been imposed at the Restoration on the admission of woollen goods into England,[6] and the colonial market was closed by the Navigation Acts. In practice, therefore, the products of the Irish woollen manufacturer had to look for markets in foreign countries.

The restraints on the export of raw wool were felt as a great hardship at the time, as wool was one of Ireland's principal products. In 1661 the Committee of Trade

[1] C.S.P., Ire., 1660-2, p. 691.   [2] C.S.P., Dom., 1677-8, p. 287.
[3] C.S.P., Ire., 1660-2, p. 590.   [4] Carte, vol. iv., pp. 287-90.
[5] Bagwell, vol. iii, p. 237.   [6] 12 Car. II., c. 4.

reported to the House of Commons that the restraints should be repealed, and Temple, who was Solicitor-General, announced that a bill to do this was under consideration.¹ Needless to say, no such bill was ever introduced, much less passed, and the matter was again agitated in 1666 after the passing of the Act restraining the import of Irish cattle into England, but no satisfaction was obtained.² In 1669 instructions were sent from England that the prohibition must be strictly enforced.³

The owners of Irish sheep had to find some means of getting rid of their great quantities of wool, and three ways presented themselves—first, sending the wool to England and paying the high licence duties; second, smuggling the wool abroad; and, third, manufacturing it into cloth in Ireland.

Great quantities were sent to England in spite of the high licence and other duties which were payable on it,⁴ and the granting of these licences was also a frequent occasion of extortion and abuse.⁵ The following fees had to be paid before the wool could even be put aboard the ship⁶:—

PER STONE.

Prime Duty	.. 15d.
License	... 4d.
Fees for License 1,500 stones ...	£1 5s. 0d.

PETTY DUTY.

Searcher	4d. per Bag.
Cranier	˙... 4d. per Bag.
Weigher and Parceler	2d. per Bag.
Writing the Entry ...	£6 the Whole Packet.
Gabridge	... 4d.

Some idea of the quantity transported may be gathered from an account of the wool licences granted in the years 1678-81, preserved amongst the Ormond Manuscripts.⁷ For example, for the three last months in the year 1678, licences were granted for 29,600 stones; in the year 1679 licences for 163,700 stones; and in the year 1680 for

---

¹ *I.C.J.*, vol. i, p. 417.   ² C.S.P., Ire., 1666-9, pp. 184-290.
³ C.S.P., Ire., 1666-9, p. 746.   ⁴ *Ormond MSS.*, n.s. vol. iii, pp. 232-7.
⁵ C.S.P., Dom., 1672, p. 628.   ⁶ *Ormond MSS.*, n.s. vol. iii, p. 335.
⁷ *Ormond MSS.*, n.s. vol. iv., p. 665.

177,600 stones. In the years immediately preceding the Revolution, about 200,000 stones of Irish wool were imported into England annually.[1]

An unlimited importation of Irish wool suited the interests of the English weavers; but at last it became apparent that it affected adversely the interests of another and more powerful section of the community, the landowners, who feared that the value of their own sheep might deteriorate.[2] But the clothing interest was anxious for a more and more unlimited importation, and a proposal was made that a company should buy all the wool produced in Ireland to be sent to the English market at a fixed price.[3]

The lists of wool licences do not furnish anything like a full account of the amount of wool exported from Ireland. A better price could be obtained in Spain, France, and Holland than in England, as these countries were unable to work up their own wool unless it had some English or Irish wool mixed with it, and were consequently willing to pay a high price for such an essential article. As early as 1660 we read a complaint of the smuggling of Irish wool,[4] and in 1663 great quantities were said to be smuggled to Holland.[5] / This clandestine exportation increased after the Cattle Act; in 1669 a great quantity of wool was sent to France packed so as to deceive the Customs Officers into the idea that the packages contained cloth.[6] " The mischief and prejudice done to the clothing trade of England by exporting wool from Ireland to foreign parts under the pretence of shipping it for England is great and manifest, and this cannot be prevented here, though all care is taken for it. For though it is felony to ship any wool out of Ireland

[1] *Remarks on Interest and Trade of England and Ireland*, London, 1691; and Petty, *Treatise of Ireland*, 1685; Roberts, *The Merchant's Map of Commerce*, London, 1677. In 1671, 352,307 stones of Wool were exported, C.S.P., Dom., 1671, p. 507.
[2] C.S.P., Dom., 1677-8, p. 37.
This proposal, which throws a good deal of light on the woollen industry of the date, is printed in Appendix x.
[4] C.S.P., Ire., 1660-2, pp. 151-2; Murray, *Commercial Relations*, p. 98.
[5] C.S.P., Ire., 1669-70, p. 476.
[6] *England's Interest Asserted in the Improvement of its Native Commodities*, London, 1669.

without a licence, and although those that have such licences do enter into bonds to return certificates from the Customs House in England that it was landed there accordingly, yet wool is carried in great quantities for France and Holland, and certificates obtained or presumed that such wool was landed in England."[1] In 1673 the evil of smuggling had grown to such an extent that it was decided in future to grant licences to export wool from six ports only, and thus to some extent centralize the control of the Customs authorities.[2] In 1676 the English clothiers declared that their continental trade was being ruined by reason of the quantity of Irish wool that found its way abroad.[3] In 1679 it was said that the French bought most of the wool growing in Ireland, and had in consequence almost ruined the English industry.[4] In one year no less than forty ships left Ireland with wool, ostensibly bound for England, but actually bound for France.[5]

One of the remedies proposed to diminish smuggling was the establishment of a wool staple. The proposers of this step argued that Ireland would gain because of the steady market for its wool, and England because all the wool produced in Ireland would go there; and furthermore it had the additional advantage that " Ireland would have no need to start a manufacture at home if a good price for wool was obtained from the stapler."[6] Luckily, this proposal was not acted upon, possibly because the staple in the earlier part of the century had not been an unqualified success.

The last, and economically most commendable way of meeting the restraints on the export of wool, was to manufacture the wool into cloth in Ireland. The tendency to do this was accentuated by the passing of the Act in 1666 preventing the importation of Irish cattle into

---
[1] *Memorandum Respecting Wool Exports*, Ormond MSS., n.s. vol. iv, p. 116.
[2] C.S.P., Dom., 1673, pp. 337, 54.   [3] C.S.P., Dom., 1676-7, pp. 217, 219; 1677-8, p. 71.
[4] *An Account of the French Usurpation Against the Trade of England*, by J. B., London, 1679.   [5] *A Plea for the Bringing in of Irish Cattle*, London, 1680.
[6] C.S.P., Ire., 1660-2, pp. 151-2.

England, which had the effect of greatly increasing the quantity of wool in the country.¹

To encourage the woollen manufacture was one of Ormond's great ambitions, and, with this end in view, he succeeded in getting passed an Act, which, after reciting that the credit of Irish woollen cloth was impaired abroad by the false and uncertain making thereof, provided that all old and new drapery should be made of specified dimensions, and that an alnage office and alnager should be appointed to regulate the inspection and sealing of the cloth.² This Act was not a success. In 1666 Sir George Rawdon referred to "the alnage business which has made such a disturbance in this and other markets by seizing every piece of cloth and woollen stuff, by the corruption and knavery of those that form it. . . . These projects are not the way to increase but to hinder our little trade."³ It does not even seem to have succeeded in reforming the abuse at which it was originally aimed; in 1682 there were still complaints of "the abuses in our worsted manufacture by which their value and credit are impaired."⁴ In 1695 a petition was presented to the House of Commons from the corporation of weavers of Dublin, on behalf of all the weavers in Ireland, complaining that the Act was in many respects destructive of the woollen manufacture, that it was not carried out properly, and that the alnager's deputies were incompetent and accustomed to demand excessive fees.⁵ In the same year the Committee of Trade reported that the alnage Act was impracticable and prejudicial, and advised that it should be repealed.⁶ The Act, however, was not repealed until 1779.⁷

The woollen manufacture was further encouraged by high duties imposed on the import of foreign woollens,⁸ and in 1667 the import of woollen goods from Scotland was prohibited.⁹

¹ *Reasons for a Limited Exportation of Wool*, London, 1677; Murray, *Commercial Relations*, p. 38.   ² 17 & 18 Car. II., c. 15.   ³ C.S.P., Ire., 1666-9, p. 5.
⁴ Lawrence, *The Interest of Ireland in its Trade and Wealth Stated*, Dublin, 1682.
⁵ *I.C.J.*, vol. ii, p. 62.   ⁶ *I.C.J.*, vol. ii, p. 95.   ⁷ 19 & 20 Geo. III., c. 20.
⁸ 14 & 15 Car. II., c. 9.   ⁹ Carte, vol. iv, p. 288.

The beginnings of a manufacture of woollen goods for export are to be traced in the petition of one Giovanni Battisti Benzie to the Irish Privy Council for assistance in his plan of sending "a person from Leydon to Limerick to employ them in making bags and such like stuffs to be retailed in Spain and Italy."[1] The industry did not, however, progress rapidly; in 1662 there was not enough cloth made in Ireland to supply even the home market, and many kinds of woollen goods were imported.[2] The greater part of the exports was made up of frieze and coarse fabrics, whereas the imports consisted almost altogether of cloth, and the finer articles of manufacture. It may be inferred from this that the Irish did not at that time manufacture much of the finer stuffs; this is also proved by the petition that the duty on English articles of clothing imported into Ireland should be lowered.[3] In 1666 we read of "a kind of cloth manufactured in the country, which is very cheap, and is carried to Spain, Italy, and often to the American Islands."[4]

The slow growth of the industry may be further inferred from a memorial which Sir Peter Pett presented to the Duke of Ormond in 1667 for the erecting of a manufacture of cloth; "for though through the want of Spanish wool, proper streams, and fuller's earth, they could not propose to carry on a trade with it abroad, yet they might make a sufficient quantity for their consumption at home."[5] The following table of exports[6] shows the slow progress which the industry was making at this time :—

	Year Ending 25th Dec., 1665.	Year Ending 25th Dec., 1669.
Wool, great stones ...	131,013	254,760
Broadcloth (pieces of 36 yards)	32	29
Stuff, pieces	224	315
Frieze, yards	444,381	392,735
Frieze Stockings (doz. pairs) ...	1,840	1,309
Rugs	321	61
Caddowes and Blankets ...	763	189

[1] C.S.P., Ire., 1660-2, p. 169.   [2] C.S.P., Ire., 1660-2, p. 584.
[3] I.C.J., vol. i, p. 416; C.S.P., Ire., 1663-5, p. 626.
[4] Jorevin de Rocheford, *Description of England and Ireland after the Restoration.*
[5] Carte, vol. iv, p. 283.   [6] C.S.P., Ire., 1669-70, p. 54.

Petty, in 1672, calculated that there were about four million sheep in Ireland.¹ The majority of the people in the country wore clothes made from the wool of their own sheep, spun into yarn by themselves;² " the clothing is a narrow sort of frieze of about twenty inches broad, whereof two foot, called a bandle, is worth 3½d. to 18d. Of this seventeen bandles make a man's suit, and twelve bandles make a cloak. According to which measures and proportions and the number of people who wear this stuff, it seems that near thrice as much is spent in Ireland as exported; whereas others have thought quite contrary, that is, that the exported wool is triple in quantity to what is spent at home."³ Petty does not say much about the woollen manufacture, and we may therefore conclude that it was not a very conspicuous feature at the time; he simply says that there were 30,000 " workers of wool and their wives " in Ireland;⁴ and that " the clothing trade is not arrived to what it was before the late Rebellion. And the art of making the excellent, thick, warm, coverlets, seems to be lost and not yet recovered."⁵

In the following year Sir William Temple described what he had observed of the wool trade and woollen industry: " The wool of Ireland seems not to be capable of any increase, nor to suffer under any defect, the country being generally full stocked with sheep, cleared of wolves, the soil little subject to other rots than of hunger; and all the considerable flocks being of English breed and the staple of wool generally equal with that of Northampton or Leicestershire, the improvement of this commodity by manufacturers in this Kingdom would give so great a danger to the trade of England that it seems not fit to be encouraged here, at least no further than to such a quanity of one or two summer stuffs, Irish frieze and cloth from six shillings to fourteen, as may supply in some measure the ordinary consumption of the King-

---

¹ *Political Anatomy*, p. 56.   ² *Ibidem, p.* 188.   ³ *Ibidem*, p. 201.
⁴ *Ibidem*, p. 146.   ⁵ *Ibidem*, p. 209.

dom. That which seems most necessary in this land is the careful and severe execution of the statutes provided to forbid the exportation of wool to any other parts but to England; which is the more to be watched and feared since thereby the present riches of this Kingdom would be mightily increased, and great advantages might be made by the connivance of Governors; whereas on the other side, this would prove a most sensible decay, if not destruction of manufactures, both here and in England itself."[1]

The success, such as it was, of the Irish woollen industry was beginning to attract a good deal of jealous attention in England. More than one English clothier urged that the industry should be made give way to a linen industry, which could not injure English interests;[2] and in 1675 the weavers of Exeter destroyed the looms of some Devonshire weavers, who, it was rumoured, intended to emigrate to Ireland.[3] About the same time the export of fuller's earth from England to Ireland was prohibited.[4]

The next few years were those in which the woollen industry made its most rapid progress. In 1674 the Duke of Ormond set up a manufacture at Callan, in connection with which undertaking Sir Richard Lawrence wrote to Captain Mathew.[5]

"We cannot make too much of this commodity if we can make it so much cheaper than England as will repay the disproportion of freight and adventure, which will be above treble what it is from London or Hull, in regard the voyage and adventure is so much more hazardous and chargeable. We have the advantage of England thirty pounds per cent. in the wools. If we can reduce the labour near the rate of England, the design will take, and then we may as well transport £5,000 worth a year as one. If we cannot attain to this foreign manufacture, it is vain to attempt any other, depending

[1] Temple, *Essay on Trade of Ireland*, 1673. [2] C.S.P., Dom., 1675-6, p. 277.
[3] C.S.P., Dom., 1675-6, p. 329. [4] C.S.P., Dom., 1677-8, p. 583.
[5] *Ormond MSS.*, vol. iii, p. 347.

upon the expense of the country. The trouble and uncertainty of sales, and hazards of trusting is not to be endured, and the country wears little of this sort of coarse cloth; and for fine cloth, our spinning will not come to it in the seven years for any quantity I concern myself in. .... If this design fail, I will never meddle more with manufacture in Ireland; but it appears to me to be the most rational and hopeful that hath been undertaken; if there be anything either in the articles or in my letter obscure, upon notice I shall explain."[1] In the same year a woollen manufactory was set up at Clonmel, also with the help of the Duke of Ormond;[2] and this manufacture was a great success, giving employment to many hundreds of workers, and making as good cloth and stuff as any that England could produce.[3] The woollen manufacture was set up in Cork and Waterford by Huguenots,[4] and several smaller enterprises were set on foot by Dutch and English immigrants.[5]

The woollen manufacture suffered to some extent when the protecting hand of Ormond was withdrawn, and the industry seems to have gone through a period of decline about 1682.[6] We also read complaints of unskilful workmanship, and of Irish goods which could not possibly be said to be as good as English goods of the same sort.[7]

It was a generally accepted truism among the Williamites after the Revolution that the Irish woollen industry declined during the reign of James II., but the following figures[8] show that this was not the case:—

WOOLLEN GOODS EXPORTED OUT OF IRELAND.

Year.	New Drapery, pieces.	Old Drapery, pieces.	Frieze.
1685	224	32	444,381
1687	11,360	103	1,129,716

[1] Much light is thrown on the nature of the Duke of Ormond's undertakings by the two agreements printed in Appendix xi. [2] *Ormond MSS.*, vol. iii, p. 353, *et. sq.*
[3] Lawrence, *The Interest of Ireland in its Trade and Wealth Stated*, London, 1682, vol. ii, p. 189; Carte, vol. ii, pp. 283-4. [4] Murray, *Commercial Relations*, pp. 99-100.
[5] A good account of these undertakings is to be found in a pamphlet entitled, *Letter from a Gentleman in Ireland to his Brother in England*, London, 1677.
[6] Lawrence, *The Interest of Ireland in its Trade and Wealth Stated*, London, 1682, vol.ii,p.190. [7] *Remarks on Trade and Interest of England and Ireland*, London,1691.
[8] Murray, *Revolutionary Ireland and its Settlement*, p. 395.

It must also be remembered that the exports are not in themselves an exclusive criterion by which to judge the growth of the industry; the population of the country increased during those years, and we have no evidence to show that there was any increased import of woollen goods to supply the Irish consumer.

On the whole, the woollen industry does not seem to have grown quite so rapidly in this period as is sometimes supposed. Even in 1677, when the industry was at its height, it was said that there was not as much wool worked up in the whole of Ireland as in a single county in England.[1] The quantities of wool exported were too great to be consistent with the existence of a really large home demand. Besides, if the Irish industry had attained very great proportions, Irish woollen goods would have undersold English goods in foreign countries, and it is quite certain that if this had occurred the industry would have suffered some hostile action from England. We should probably be correct in concluding that, so long as there were no signs of jealousy in England, the Irish woollen industry was of no great magnitude; and, as we shall see in a later section, jealousy of the Irish woollen industry first sprang into dangerous prominence in the years immediately following the Revolution.[2]

(12) *The Linen Industry.*

As we have seen, the small linen trade which existed in Ireland in the early seventeenth century was ruined during the Rebellion, and, as in the case of the woollen industry, a fresh start had to be made after the Restoration. The industry was encouraged by the Irish Parliament; in 1662 the Committee of Trade was instructed to consider how the linen manufacture might be encouraged and advanced; and this Committee's report was referred

[1] *Letter from a Gentleman in Ireland to his Brother in England Relating to Trade*, London, 1677.
[2] Miss Murray also concludes that it was "After the Revolution that Irish Woollen Manufactures began to make really rapid strides," *Commercial Relations*. p. 102.

to a sub-committee to be embodied in a bill.¹ It was probably the report of this Committee which formed the basis of the Act, passed a couple of years later, which provided (1) that nobody should let a cottage or cabin to anybody not holding one acre plantation measure, and that the tenant of such a cabin should sow one-eighth of his holding with hemp or flax; (2) that persons who ploughed any land should sow half an acre with hemp or flax for every thirty acres ploughed; (3) that no linen cloth should be woven, bought, or sold less than three-quarters of a yard broad out of the loom; (4) that each county in Ireland should raise twenty pounds annually for the next twenty years, and award prizes to the weavers of the three best pieces of linen each year; (5) that six thousand pounds should be raised to purchase a bleach-house of four acres in each province, and that tenements should be erected for poor people and idle vagrants who were to be set to work at the linen manufacture; and (6) that weavers, who did not at the same time exercise any other trade, should be exempted for seven years from serving on juries or bearing offices.²

This Act was not enforced with sufficient vigour to make it effective. In 1672 Sir William Petty complained that flax and hemp were not being sown in accordance with the statute,³ and in 1680 Ormond pronounced the collection of the penalties for non-fulfilment of the Act as "impracticable, but with the highest discontent of the people of all sorts."⁴

When this Act was passed, the linen industry was still in a very undeveloped state. "We sent abroad little linen cloth and less that is good, though store of linen yarn, which is an imperfect sort of country manufacture, and showeth that we have more spinners than weavers."⁵ The exports and imports of linen for the year 1665⁶ were as follows :—

[1] *I.C.J.*, vol. i, pp. 571, 586.   [2] 17 & 18 Car. II., c. 9.
[3] *Political Anatomy*, p. 122.   [4] *Ormond MSS.*, vol. v, p. 421.
[5] C.S.P., Ire., 1662-6, pp. 693-8.   [6] C.S.P., Ire., 1662-6, p. 698.

EXPORTS.

		To Foreign parts.	To England.	Total.
Linen Cloth,	ells.	5840	5960	11,800
Linen Yarn,	cwt.	—	3477	3,477
IMPORTS.				
		From England.	From Foreign parts.	Total.
Pieces,	value £	9526	2654	12,180

The industry commenced to progress about the year 1666, but this was probably caused, not so much by the Act of Parliament of that year, as by the active encouragement it received from the Duke of Ormond. " The Duke of Ormond, as soon as he came over into Ireland, undertook the revival of this manufacture and got Acts of Parliament passed for the encouragement of it, and for inviting Protestant strangers to settle in the Kingdom. He was at the charge of sending understanding persons into the Low Countries to make observations on the state of trade in those parts, their manner of working, the way of whitening their thread, the laws and statutes by which the manufacture was regulated, the management of their grounds for hemp and flax, and to contract with some of their most experienced artists. He engaged Sir W. Temple to send him over out of Brabant five hundred families that had been employed in that manufacture; he procured others from Rochelle and the Isle of Rhé, and Sir G. Carteret supplied him with a considerable number from Jersey, and the neighbouring parts of France. He built tenements for the reception of many of these at Chapelizod, near Dublin, where, before he went the next year to England, there were three hundred hands at work in making cordage, sail-cloth, ticking, and as good Irish linen and diaper of Irish yarn as was made in any country of Europe. This was carried on under the direction of Colonel R. Lawrence. His Grace erected another manufacture of this sort in his own town of Carrick, assigning to the workmen half the houses in the place and five hundred acres of land. It will not give one an advantageous opinion of the industry of the people of Ireland to find that the want of spinners and the ignorance of that art were the greatest obstructions met with in carrying the several branches of the linen manufacture to perfection. But the Duke of Ormond's example and encouragement got over all difficulties, and improved it in a very few years to such a degree that the whole nation felt the benefit thereof, and it is now the most considerable part of its commerce. The difficulties, hazards, and expense were great at the beginning of this work; but they were all surmounted before he left the Government in 1669, when the Nation was in such a way of flourishing with these manufactures that Colonel Lawrence did not question but posterity would own their future affluence to be a blessing they derived from his Grace's wisdom and excellent government."[1]

At the same time, Lord Orrery founded a linen manufactory at Charleville. "I am very busy," he writes in 1667, "settling a linen manufacture in this place. I have laid out no inconsiderable sum in building a whole street for all sorts of linen manufacture, as also in providing artists, looms, and all sorts of instruments, the best sorts of Flanders and Brabant seeds, and in making a magazine of victuals for a great number of apprentices whereby not only to carry on the manufacture here, but in seven years to disperse it all over the province."[2]

Sir William Temple gave the following account of the industry in 1673:—" Yarn is a commodity very proper to this country, but made in no great quantities in any parts except the north, nor anywhere into linen to any great degree, or of sorts fit for the better uses at home, or exportation abroad; though of all others this ought most to be

[1] Carte, vol. iv, pp. 284-6.   [2] C.S.P., Ire., 1666-9, p. 367.

encouraged. .... This may certainly be advanced and improved into a great manufacture of linen, so as to beat down the trade both of France and Holland, and draw much of the money which goes from England to those parts upon this occasion into the hands of his Majesty's subjects in Ireland, *without crossing any interest of trade in England.* Much care was spent upon this design in the Act of Parliament passed the last session, and something may have been advanced by it; but the too great rigour imposed upon the sowing of certain quantities of flax has caused a general neglect in the execution; and common guilt has made the penalties impracticable; so as the main effect has been spoiled by too much diligence, and the child killed with kindness. For the money applied by that Act to the encouragement of making fine linen, though the institution was good, yet I think it has not reached the end by encouraging any considerable application that way; so that sometimes one share of that money is paid to a single pretender at the Assizes or Sessions, and sometimes a share is saved for want of any pretender at all.

" There are but two things which can make any extraordinary advance in this branch of trade—first, an increase of the people in the country to such a degree as may make many things necessary to life dear, and thereby force industry from each member of a family; the second is a particular application in the government.'"[1]

In 1678 it was suggested that some of the taxes payable in Ireland should be paid in flax or hemp, and the English Government showed a disposition to encourage the industry in other ways.[2] Lord Chancellor Boyle, however, who was consulted on the matter, thought no further encouragement was needed, but that the existing Act of Parliament would be sufficient if it were properly carried out: " I am of opinion that, by a prudent management of the Act of Parliament which we have already in this Kingdom upon that account, a very considerable benefit

---

[1] Temple, *Essay on Trade of Ireland*, 1673.   [2] *Ormond MSS.*, vol. iv., p. 175.

may arise thereby both to the King and Kingdom. Nothing in my opinion hath given a greater discouragement to the linen manufacture than Colonel Lawrence's failure in his works at Chapelizod. Either he must not have understood the trade, or else must be a gross hypocrite, for from his failure the argument is thus drawn to the disadvantage of that manufacture. If Colonel Lawrence could not support that small undertaking with all the advance he had of money from the Government, and the continued help of the taking of his linen in very great quantities for the use of the army, how can it be expected that any great or considerable benefit may be raised by that manufacture? But notwithstanding the force of this objection I must yet believe that exceeding much more may be made by the linen trade in this Kingdom than hath yet been done, and very considerable advantage by the cordage of the hemp."[1] The bleach-yards and weaving-shops at Chapelizod were afterwards granted for twenty-one years to one Christopher Lovett, together with a sum of £1,200.[2]

In spite of the failure of the manufacture at Chapelizod, the industry as a whole continued to progress. "The Scotch and Irish in Ulster," wrote Lawrence in 1682,[3] " addicting themselves to spinning of linen yarn, attained to vast quantities of that commodity, which they transported to their great profit, the convenience of which drew thither multitudes of linen weavers, and my opinion is that there is not a greater quantity of linen produced in the like circuit in Europe, and although the generality of their cloth fourteen years since was *sleisie* and thin, yet of late it has much improved to a great fineness and strength." In 1685, according to Petty, great quantities of linen cloth were exported.[4]

---

[1] *Ormond MSS.*, vol. iv, pp. 185-6. [2] C.S.P., Dom., 1690-1, p. 338.
[3] *The Interest of Ireland in its Trade and Wealth Stated*, vol. ii, p. 190.
[4] Petty, *Treatise of Ireland*; *The Appeal of the Protestants of Ireland*, London, 1689.

(13) *The Fishing Industry.*

As we have seen, the fishing industry was completely destroyed during the Rebellion, and it does not seem to have attained any improvement for many years after the Restoration. The exports of fish in 1665 and in 1669 were small[1]:—

Fish.		1665.	1669.
Herring	barrels.	16,252	12,893
Salmon	tons.	330	905
Pilchards	,,	332	795
Hake	no.	178	560
Train Oil	tuns.	26	107

In 1672 Petty stated that there were only one thousand people in the whole of Ireland engaged in fishing.[2] "There are in the West of Ireland about twenty gentlemen who have engaged in the pilchard fishing, and have among them about 160 saynes wherewith they sometimes take about four thousand hogsheads of pilchards worth about ten thousand pounds. Cork, Kinsale and Bantry are the best places for eating of fresh fish, though Dublin be not or need not be ill-supplied with the same."[3] In the following year Sir William Temple complained that the fisheries were very unprofitable; "the fish of Ireland might prove a mine under water as much as any under ground, if it was improved to those vast advantages it is capable of, but that is impossible under so great a want of people and the cheapness of things necessary to life throughout the country."[4] The absence of the fishing industry on any considerable scale may be gathered by the fact that, amongst the industries mentioned which were stated to be likely to suffer from a tax on salt, there is no mention made of the fisheries.[5]

Various attempts were made to improve the Irish fisheries during this period. In 1663 an Order in Council

[1] C.S.P., Ire., 1669-70, p. 54.   [2] *Political Anatomy*, p. 12.   [3] *Ibid.*, p. 111.
[4] *Essay on the Trade of Ireland*, 1673.   [5] C.S.P., Ire., 1669-70, p. 321.

provided that no foreigner should fish in Irish waters during the pilchard season, but this regulation was not carried out;[1] three years later Ormond attempted to erect a company to promote fishing, but did not succeed in doing so;[2] and in the following year Sir George Rawdon endeavoured to provide boats and nets for Irish fishermen before the herring season opened, so that Scotch boats might be prevented from deriving all the profit from the Irish waters.[3] In 1672 some doggers captured from the Dutch were sent to Ireland to help in setting up a fishery.[4]

The successful development of this industry was hampered by the competition of the French fishermen, who came in great numbers and resisted by armed force all attempts to remove them. We read complaints of this invasion taking place off the south of Cork in 1670,[5] and in the following year the Lord Lieutenant expressed himself quite unable to cope with it, and thought it was better " rather to suffer the inconvenience from fishing on the coast than by one over hasty order to expose our men to receive affront, we having no ship on the coast to make good against so many boats armed as they are.'"[6] The following letter written from Kinsale gives a good idea of the extent to which this evil had grown :—" About thirty or forty French fishing boats are come on this coast, each of thirty or forty tons burden, having very long strings and rafts of nets, which they call mackarel nets. But they have both mackarel and herring nets; the mackarel they place uppermost next the ropes and the herring under the same raft and joined together. Each having 100 nets and about 30 or 40 men carries out in length about two miles. All their men are armed with muskets and fire locks. By these long and unlawful nets they break and destroy the great shoals of fish on this coast, to the great destruction of the fishing trade here and particularly of the pilchard fishing in the West of Ireland, set forth at very great charges of the undertakers and of the hookers

---

[1] C.S.P., Dom., 1671, p. 196.   [2] Carte, vol. iv, p. 282.   [3] C.S.P., Ire., 1666-9, p. 341.
[4] C.S.P., Dom., 1672-3, p. 31.   [5] C.S.P., Ire., 1669-70, p. 132.   [6] C.S.P., Dom., 1671, p. 196.

and fishermen of this town, consisting of 60 or 80 boats, and also those of Youghal and Dungarvan, to their great disheartening and impoverishment. Before these French vessels came it was very usual for the hookers and fishermen of Kinsale, with about three men and a boy in each boat, to take 3,000 or 4,000 mackarel a day; but now they take few or none, and also other fish grow very scarce."[1] The fisheries were also impeded by the difficulty of getting salt, and frequently fish had to be used for manure, as there were no means available of curing it.[2]

The fisheries seem to have improved towards the end of the century. It was said in 1689 that " the cargoes of salmon, herrings, and pilchards and other fish made up yearly in Ireland and transported into several ports of Spain and Venice and of the ports in the Mediterranean Sea would startle common people."[3]

(14) *Minor Industries.*

The glass industry did not exist in any appreciable degree during this period. In 1670 a small glassworks was set up at Portarlington; a good deal of money was laid out on the manufacture; and several families went to live there in consequence.[4] About the same time there were several glassmakers in Dublin,[5] but on the whole it may be said that the period of the prosperous Irish glass industry had not yet begun. The pottery industry also existed only on a very small scale. In 1667 a monopoly was granted for " making pantiles, paving tiles, and other sorts of tiles, as also for making pots, dishes and all other earthenware never found out, used, or invented before in Ireland ";[6] and there is some evidence of the existence of a pottery at Belfast in 1688.[7]

The brewing industry continued to flourish, as it had done before the Rebellion. In 1670 the brewers of Dublin

---
[1] C.S.P., Dom., 1671, pp. 196-7.
[2] *The Interest of England in the Preservation of Ireland,* by G. P., London, 1689.
[3] *Ibidem.*   [4] C.S.P., Ire., 1669-70, pp. 301-2.
[5] Westropp, *Glassmaking in Ireland,* Proc. R.I.A., vol. xxix. c., p. 34.
[6] C.S.P., Ire., 1666-9, p. 285.
[7] Westropp, *Notes on Pottery Manufacture in Ireland,* Proc. R.I.A., vol. xxxiii. c., p. 2.

were granted a charter.¹ In 1672 Petty wrote: "In Dublin where are but 4,000 families, there are at one time 1,180 ale houses and 91 public brew houses";² and in the same year an English traveller referred to "the incomparable beer and ale, which runs as freely as water on a visit."³ The brewers must have been a powerful and well-organized body; in 1676 they petitioned Parliament about the oppressive way in which the excise was collected;⁴ and this was the beginning of a long conflict, which it would not have been worth the Government's while to engage in, were not the revenue involved considerable.⁵ "By a reasonable addition to the excise of beer and ale," wrote the Duke of Ormond, "twenty thousand pounds a year will come to the King."⁶ This shows that the industry must have been very considerable, as the excise payable on beer and ale was only 2s. 6d. a barrel when the price was more than six shillings a barrel, and 6d. a barrel when under that price.⁷ In 1679 a correspondent of the Duke of Ormond wrote from the Hague suggesting that "a company of brewers" should be sent from Holland to Carrick-on-Suir to start a brewery there.⁸ In 1685, 3,541 hundredweight of hops were imported into Ireland;⁹ and in the same year 4,644 barrels of beer were exported.¹⁰ In addition to the public brewers, a great many people brewed ale on a small scale in their own little taverns.¹¹

It is doubtful whether shipbuilding was carried on anywhere in Ireland during these years except at Belfast, where it existed on a very small scale.¹² Another industry which existed on a small scale was that of sugar refining; this was introduced into Ireland for the first time in 1668,¹³

---
¹ C.S.P., Ire., 1669-70, p. 277.   ² *Political Anatomy*, p. 13.
³ *A Tour in Ireland in* 1672-4, Journal of Cork Arch. Soc., vol. x, p. 90.
⁴ *I.C.J.*, vol. ii, p. 20.   ⁵ *Ormond MSS.*, vol. iv, p. 163; vol. vii, p. 98.
⁶ *Ormond MSS.*, vol. iv, p. 112.   ⁷ 14 & 15 Car. II, c. 8.
⁸ *Ormond MSS.*, vol. v, p. 245; cider was also made in Ireland, C.S.P., Ire., 1666-9, pp. 473, 531.
⁹ *Remarks on the Affairs and Trade of England and Ireland*, London, 1691.
¹⁰ Petty, *Treatise of Ireland*.
¹¹ *A New Irish Prognostication*, London, 1689. The existence of a large brewing industry may be inferred from the elaborate provisions made in 1683 for the collection of the excise on beer in that year. These are to be found in a book contained in Box 89 of the Halliday Collection of Unbound Tracts in the Royal Irish Academy.
¹² Benn, *History of Belfast*, p. 310.   ¹³ C.S.P., Ire., 1666-9, p. 561.

and we hear of its being carried on in Belfast at a later date.[1] As in earlier times, the tanning industry did not reach the proportions which it should have. In 1665, 106,344 raw hides were exported from Ireland; in 1669 the number was doubled;[2] and Petty complained of the vast quantities of untanned hides which were exported in 1685, " because the English for their paucity, and the Irish for their want of skill, could not manufacture them to the best advantage."[3] However, the industry was by no means non-existent; in 1672 Petty calculated that there were in Ireland " 20,000 shoemakers and their wives, and 10,000 tanners and curriers and their wives ";[4] in 1685, 86,013 tanned hides were exported.[5] Pewter was manufactured to some extent; from 1656 a good deal was made in Cork;[6] and the manufacture was of sufficient magnitude in 1699 to call for regulation by Act of Parliament.[7] Oilmills, probably for the manufacture of rape oil, also existed.[8] There were also manufactures of wax, furs, salt and other commodities, but only on a very small scale[9]

(15) *Public Finance.*

After the Restoration the taxes imposed by Cromwell were continued for a short time, but within a few years the whole Irish revenue was reformed and put on a new basis by statutes, which, in lieu of certain rights surrendered by the King, conferred upon him and his successors certain duties and taxes which came to be known as the hereditary revenue. At the same time the King was allowed to retain some of the old duties which had been his before the Restoration. The Irish revenue was thus composed of two distinct parts, first, the ancient patrimony of the Crown payable by prescription or custom sanctioned at common law, such as Crown rents, port corn and composition rents, prizage, lighthouse duties and casual revenue; and second, the duties granted to

[1] Benn, *History of Belfast*, p. 332.   [2] C.S.P., Ire., 1661-70, p. 54.
[3] *Treatise of Ireland.*   [4] *Political Anatomy*, p. 13.   [5] *Treatise of Ireland.*
[6] *Journal of the Cork Arch. Soc.*, vol. ix, p. 32.   [7] 9 Will. III, c. 14.
[8] *Dinely's Tour*, Jl. of Kilkenny Arch. Soc., vol. v., n.s. p. 43.
[9] Roberts, *The Merchants' Map of Commerce*, London, 1677.

Charles II. by Parliament in exchange for branches of the ancient revenue of the Crown that had been found grievous to the subject, such as wardship and feudal dues, or in return for forfeitures. This second branch of the revenue included quit rents, customs inward and outward, inland and import excise, fines, seizures and forfeitures, licences for sale of beer, ale and spirits, and hearth money.

The quit rents arose from the forfeitures of the Rebellion of 1641, and were fixed by the Act of Settlement to be paid out of forfeited lands at a certain amount per acre—in Leinster 3d., in Munster 2¼d., in Ulster 2d., and in Connacht 1½d.[1] The customs were composed of duties of poundage and tonnage. Poundage was composed of two separate duties, the old poundage and the new poundage, the old poundage consisting of a duty of twelvepence in the pound on all imports and exports except wine and oil,[2] and the new poundage of a duty of the same amount on all goods exported by merchants, strangers or aliens.[3] Tonnage was a duty payable upon all wine and oils imported. The excise was divided into import and inland excise. The former was a duty on all commodities imported except jewels, bullion, coin, arms and ammunition, according to the book of rates annexed to the statutes creating it. The rate payable on drugs was two shillings in the pound; on raw hemp, flax, undressed resin, pitch, tow, wax, cables, cable yards and cordage sixpence in the pound; and on wine, tobacco and salt a shilling in the pound.[4] The inland excise was a duty payable on beer, ale and spirits at the rate of 2s. 6d. per barrel on ale and beer above 6s. in value, 6d. a barrel on ale and beer under that value, and 4d. a gallon on spirits.[5] The licence duties consisted of a tax of twenty shillings a year for retailing ale and beer and a tax on a sliding scale for retailing spirits;[6] and the fines, seizures and forfeits included in the hereditary revenue were made

---

[1] 14 Car. II, Sess. iv, c. 2; 17 & 18 Car. II, Sess. v, c. 2. [2] 15 Henry VII.
[3] 14 & 15 Car. II, c. 9. [4] 14 & 15 Car. II, c. 8. [5] 14 & 15 Car. II, c. 8.
[6] 14 & 15 Car. II, c. 6; 17 & 18 Car. II, c. 19.

up of half of all seizures condemned and sold under the acts of tonnage, poundage or import excise. The hearth money was a tax of two shillings on every fire hearth, but was not payable by any person living upon alms, or who could not earn a livelihood by labour, or by widows living in a house of the value of less than eight shillings a year who did not possess more than four pounds worth of property.[1] Another duty which should be mentioned here was the alnage duty which was created by the statute regulating the woollen trade.[2] This duty did not add to the revenue, as it was granted in perpetuity to private persons.[3]

The non-statutory part of the hereditary revenue was composed of prizage, lighthouse duties, wool licences and casualties. Prizage was a prescriptive right which the Crown had to a certain amount of wine out of every ship landing wine in Ireland. The lighthouse duties were made up by a payment of fourpence per ton by every foreign ship coming to an Irish port. These duties were usually granted to a subject in consideration of building lighthouses. An attempt was made by Charles II. to exact lighthouse duties from his own subjects, but this was successfully resisted.[4] The wool licence duty consisted of the fees paid for permission to export wool, and, as we have seen, was a very lucrative branch of the revenue in the time of Charles II. The casualties were made up of a number of small miscellaneous duties—fines, forfeited recognizances, profits of the Hanaper, custodium rents and first fruits, twentieth parts, profits of faculties, waifs, estrays, felons' goods, deodands, wrecks, treasure trove, and gold and silver mines.

When the hereditary revenue proved insufficient it was supplemented by subsidies. In 1666 Parliament granted twelve subsidies of £15,000 each;[5] and three years later granted eight further subsidies of the same amount.[6] These subsidies were to be levied at the same rate as the

---

[1] 14 & 15 Car. II, c. 17 ; 17 & 18 Car. II, c. 18.   [2] 17 & 18 Car. II, c. 15.
[3] Clarendon, *Revenue of Ireland*, p. 25.   [4] Clarendon, *op. cit.*, p. 20.
[5] 14 & 15 Car. II, c. 6 and c. 7.   [6] 17 & 18 Car. II, c. 1 and c. 17.

subsidies granted in earlier reigns, which we have described in a previous chapter, and it is interesting to note that some of them were allowed to be paid, not in money, but in wheat or oatmeal.[1] Deficits in the revenue were also made good by borrowing. In 1670 there was a debt of £245,000;[2] and three years later the debt still amounted to a considerable sum.[3]

The increasing amount produced by these taxes during the period under review shows that the country increased in prosperity during that period. The following is a return of the revenue in 1663-4[4]:—

YEAR ENDING 20th MARCH, 1664.

	£	s.	d.
Old Crown Rents	9,361	10	0
New Rents	467	5	6
Composition Rents	351	13	6
New Quit Rents	38,042	5	1
Rents of Impropriate Rectories	109	11	3
Excise and Customs	55,490	18	6
Inland Excise	30,274	15	9
Licenses for Beer, Ale, &c.	3,379	0	0
Green War Money.	3,306	6	3
Felon's Goods	51	10	8
Hanaper Office	240	15	0
First Fruits and Twentieth Parts	493	2	6
Hearth Money	11,637	19	8
	£153,206	13	8

Five years later the revenue was farmed for £219,500,[5] and in 1664 it was stated to have risen to £300,000.

The following is a full return of the revenue for three years 1683-5[7]:—

	Year ending Oct. 25, 1683	Year ending Dec. 25, 1684	Year ending Dec. 25, 1685
Customs Inward and Exported Excise	85,844	91,424	91,117
Customs Outwards	32,092	33,425	29,428
Seizures	965	615	460
Prisage	1,452	1,693	1,882
Inland Excise	18,344	77,580	79,169
Ale Licenses	8,283	9,538	9,995
Wine and Strong Water Licenses	2,736	3,114	3,467
Quit Crown and Custodium Rents	68,699	68,385	68,922
Hearth Money	31,041	31,646	32,953
Casual Revenue	820	1,745	1,564
	300,279	319,368	318,961

[1] C.S.P., Ire., 1666-9, pp. 296-7. [2] C.S.P., Dom., 1671, p. 54. [3] C.S.P., Dom., 1673, p. 554.
[4] C.S.P., Ire., 1663-5, p. 555. [5] C.S.P., Ire., 1666-9, p. 767.
[6] C.S.P., Dom., 1673-5, p. 159; see Murray's *Commercial Relations*, p. 154.
[7] MS. in Marsh's Library, Z-2-17, No. 21.

During this period the revenue, on the whole, was equal to, or more than, the expenditure, and consequently there was no necessity to call a Parliament.[1] In the earlier years of the reign there were deficits, but, as we have seen, these were made good by subsidies, and by borrowing; but in 1670 the revenue and expenditure almost balanced, the former being £208,000, and the latter £210,642.[2] In 1675 there was a surplus;[3] in 1680 the revenue exceeded the expenditure by about £60,000;[4] and in 1683 there was a large surplus.[5]

The general practice of farming the revenue caused much injustice and oppression. In the first place, the farmer received a great deal more than he paid to the Government, and therefore made a large profit out of public needs; for instance, in 1680, Lord Ranelagh, who then farmed the whole revenue, received one hundred thousand pounds more than he paid into the treasury.[6] Moreover, the system of farming led to oppression of the poor, as the farmers' collectors themselves endeavoured to make a profit. " I know one of Lord Ranelagh's collectors is now reputed to be worth five thousand pounds who before his employ was not worth five groats. I as well know that the multitude of other vermin employed by him get much more, not only the commissioners, but their collectors, and every vagabond employed by them. When the vast authority of the law is without security or consideration put into the hands of vile persons for many years your Lordship may easily conjecture what terrible effects the conjunction of this authority with these persons must have on a person's property; which power being derived in its first original from obscure and impure hands like a polluted fountain, the farther it ran the more pollution is contracted, till at last it fall into the rascality of the common people and they usually exacting it with the greater rigour on their captives that Turk-like (Christian comparison the matter

[1] Clarendon, *op. cit.*, p. 25; Murray, *Commercial Relations*, p. 153.
[2] C.S.P., Ire., 1659-70, p. 339.   [3] C.S.P., Dom., 1675-6. p. 481 ; 1677-8, p. 520.
[4] C.S.P., Dom., 1679-80, p. 612.   [5] *Ormond MSS.*, vol. vii., p. 167.
[6] *Ormond MSS.*, vol. v, p. 411.

will not bear) that they may quicken and heighten the price of redemption to their own advantage."[1] "The practice of farming hath been a great trade in Ireland but a calamity on the people."[2] The poor were particularly oppressed by the hearth money which had been created in place of the revenue derived from the Court of Wards, which was essentially a revenue paid by the rich. Thus, one of the results of the Restoration was to shift the burden of taxation from the rich to the poor.[3] The evils of the system of farming had become so great and so apparent that the whole system was abolished in 1683, and the revenue after that date was collected by commissioners appointed by the Crown.[4]

On the whole, it seems clear that Ireland was heavily taxed in proportion to its wealth. In 1666 we read a complaint, which reminds one of the present day, that, whereas the wealth of Ireland was not more than one-fifteenth of that of England, the taxes which it paid were one-sixth of those paid in England;[5] and in 1683 the impositions of further taxes was advised against by the Irish Government on the ground that the country could not without bankruptcy bear any heavier taxation.[6] "The people are generally strained in their fortunes, behind-hand, and poor, and make very hard shifts to pay the present duties, insomuch as the collectors are forced many times to distrain their riding nags, milch cows, plough beasts, nay, their pots and pans and their bed clothes, which the people with great difficulty, if at all, recover, and they are often sold at near half the value, and the people are therefore fitter to receive, or at least to be forborne, than to give more."[7]

Of course, the revenue, when paid in, was at the absolute disposal of the Government, and there was no system of Parliamentary check. Theoretically, the revenue was applicable to public purposes, but in practice it

---
[1] *Ormond MSS.*, vol. v, p. 442; *Ibid.* vol. vi, pp. 32-434.
[2] Petty, *Political Anatomy*, p. 89.   [3] Carte, vol. iv, pp. 97 and 539.
[4] Carte, vol. iv, p. 641.   [5] C.S.P., Ire., 1666-9, p. 262.   [6] *Ormond MSS.*, vol. vii, p. 98.
[7] C.S.P., Dom., 1679-80, pp. 612-13.

was used by the Government in whatever way it desired. The practice, which afterwards grew to such dangerous proportions, of granting pensions payable on the Irish establishment, first began to be noticed in the reign of Charles II., and Clarendon says that at the end of that reign it had developed into a scandal.[1] In 1666, pensions and annuities amounting to £2,500 a year were paid,[2] and in 1679 they had increased to a much larger sum.[3]

Even when there was a surplus it was used by the King for his own purposes without any regard to the public need. For instance, in 1684 there was a large surplus, and the only question between the Irish and English Governments was whether it should be remitted to England or spent on the decoration of Dublin Castle; it was not at any time suggested on either side that it should be devoted to the development of Irish industry or trade.[4]

The country lost a good deal through the amount of public money which in one way or another went abroad. A certain number of Government servants were in the habit of living in England, as also were many pensioners whose incomes were charged on the Irish establishment, and about five thousand pounds a year left the country in this way for which no return was received.[5] This grievance grew so great that an attempt was made to revive the old statute against absentee office-holders, but without success.[6] As a general rule, any surplus which remained over at the end of the year was sent to England as a matter of course,[7] and Ireland was thus further drained.

(16) *Coinage and Credit.*

At the Restoration the coinage of Ireland was in the dreadful condition which we have described as existing under the Commonwealth. The country was drained of its reputable currency; the only coins in circulation were debased and "outlandish," perus and pieces of eight,

---

[1] Clarendon, *op. cit.*, p. 18. [2] C.S.P., Ire., 1666-9, pp. 78.
[3] C.S.P., Dom., 1671, p. 441. [4] *Ormond MSS.*, vol. vii, p. 178.
[5] Lawrence, *The State of Ireland, its Trade and Wealth Stated*, London, 1682.
[6] C.S.P., Dom., 1678, p. 237. [7] Macartney, *Account of Ireland*.

frequently clipped and counterfeited; and the ordinary transactions of exchange were effected through the medium of tradesmen's tokens, which had but a local circulation, and the value of which depended upon the solvency and honesty of their issuers.

The new Government made a determined attempt to reform the currency. In the first place, provision was made for the issue of a satisfactory copper coinage. To attain this object a patent was granted to Sir Thomas Armstrong, giving him the right for twenty-one years to coin copper farthings, but this patent was not made much use of, and the issue of farthings made by virtue of it was but small.[1] The general and widespread issue of private tokens was considered detrimental to the King's patentee, and the further manufacture of such tokens was consequently forbidden.[2] In spite of this prohibition, however, great numbers of tradesmen continued to issue these tokens as before, and indeed such an issue was rendered imperative by the neglect of Sir Thomas Armstrong to perform the duty which he had undertaken. Further proclamations were therefore made prohibiting these tokens in 1672 and 1673,[3] but they did not succeed in putting an end to the evil at which they were aimed.[4] On the whole, the policy of the Government in endeavouring to suppress these private issues was wise; not only did they depend for their validity on the solvency of a private individual, but local credit was liable to receive a severe shock if some malicious rival or enemy of the issuer suddenly presented for payment a large number of tokens at the same time.[5]

The Government also turned its attention to the foreign currency circulating in the country, and issued a proclamation fixing the standard weight and value of the most common foreign coins as follows :—

---

[1] Simon, *Irish Coins*, pp. 49-50. [2] Simon, p. 51. [3] Simon, pp. 52-3.
[4] *Kilkenny Archæological Journal*, n.s. vol. ii, p. 222; Lindsay, *Coinage of Ireland*, Appendix 3; Proc. R.I.A., vol. iv, Appendix 4.
[5] *Ormond MSS.*, vol. v, p. 206.

Coin.	Weight.		Value.		
	Dwt.	Grns.	£	s.	d.
GOLD.					
Rider	6	12	1	2	6
Spanish or French Quadruple Pistole	17	8	3	4	0
Double Ducat	4	12	0	18	0
Spanish Suffrain	7	2	1	8	0
SILVER.					
Mexico or Sevil Piece of Eight ...	17	0	0	4	9
Rix Dollar or Cross Dollar ...	8	12	0	2	4½
Portugal Royal	14	0	0	3	8
Ducatoon	20	16	0	5	9
Old Peru Piece and French Lewis ...	17	0	0	4	6

It was provided that twopence should be allowed off the value for every grain deficient in the weight of gold, and threepence for every penny-weight of silver.[1]

The establishment of a mint was also very seriously considered, and in 1661 a project of this kind was discussed.[2] There was a serious dispute as to whether the mint should be in public or private hands, and one party in the Government was in favour of granting a patent to three private gentlemen enabling them to coin silver pieces. Such a patent was actually issued, but occasioned so much dissatisfaction in England that the whole question was reopened and referred to a special commission. The officials of the English mint reported that such an issue would tend to debase the English currency, and the patent was, therefore, not granted.[3] The Government decided, however, to erect a public mint, and precise instructions to this effect were given immediately.[4] No mint, however, was erected, and we find that the instructions were repeated in 1671.[5] But again the Irish Government ignored the King's orders. While provision was made by law for the supply of both silver and copper coins in Ireland, no such supply was, in fact, created, and complaints of the scarcity of coin were very common throughout the remainder of the reign of Charles II. The

[1] Simon, pp. 50-1. [2] C.S.P., Ire., 1660-2, p. 109.
[3] Simon, p. 51; C.S.P., Ire., 1660-2, p. xlix.
[4] C.S.P., Ire., 1660-2. p. 544. [5] C.S.P., Dom., 1671, p. 451.

scarcity was so marked in the years immediately following the Restoration[1] that the Commonwealth money was declared to be still legal tender in Ireland even after it had been withdrawn in England.[2] In 1666 we read more complaints of shortage of currency,[3] and again in 1672.[4]

The condition of the coinage was rendered still more unsatisfactory by the prevalence of counterfeiting,[5] and by the frauds which were practised in weighing the foreign coins, whose value depended entirely on their weight. "The scales and weights differ so much from each other that what is 4s. 9d. in one house is but 4s. 6d. in the next."[6] This evil was remedied to some extent by the regulation of the manufacture of the weights used for weighing money. For instance, in 1680 a monopoly of making such weights for use in Dublin was granted by the Dublin Corporation,[7] and this probably helped in some measure to restore public confidence.[8] The most important consequence of this scarcity of money, and of the lack of confidence which the public felt in the currency, was that it was impossible to borrow money in Ireland at less than ten per cent., which had the effect of impeding trade and business very seriously.[9]

The causes of the great scarcity of coin were not far to seek. In addition to the fact that there was no mint, and therefore no constant and regular supply of money, great quantities of coin were continually being exported. The balance of trade between England and Ireland was, in the language of those days, "against Ireland"; and, in addition, large sums were remitted to England annually to pay the rents of absentees, and for the upkeep of part of the Irish army which was maintained in England.[10] The rate of exchange between the two countries was consequently very high—sometimes as high as ten per cent.[11]—

---

[1] C.S.P., Ire., 1660-2, p. 109.   [2] C.S.P., Ire., 1660-2, p. 481.
[3] C.S.P., Ire., 1666-9, pp. 10-16.   [4] C.S.P., Dom., 1672, p. 90.
[5] C.S.P., Ire., 1663-5, p. 232; 1666-9, p. 643; 1669-70, p. 536.
[6] Petty, *Political Anatomy*, p. 70.
[7] Westropp, *Money Weights and Foreign Coin in Ireland*, Proc. R.I.A., vol. xxxiii. c.
[8] Dinely, *Tour in Ireland in* 1684.
[9] Petty, *Political Anatomy*, p. 75; Lawrence, *The Interest of Ireland in its Trade and Wealth Stated*, London, 1682.
[10] C.S.P., Dom., 1673, p. 264; Carte, vol. iv, p. 261.   [11] C.S.P., Dom., 1671, pp. 328, 383, 574.

IN THE SEVENTEENTH CENTURY. 207

and consequently coin was sent in large amounts to England. Petty gives a good account of this matter:—

"But Money, that is to say, Silver and Gold, do at this day much decrease in Ireland, for the following Reasons:

"1. Ireland, Anno 1664, did not export to a much greater Value than it imported, viz., about £62,000. Since which time there hath been a Law made to prohibit the Importation of great Cattel and Sheep, alive or dead, into England; the value whereof carried into England in that very year 1664 was above £150,000. The which was said to have been done, for that Ireland drained away the Money of England. Whereas in that very year England sent to Ireland, but £91,000 less than it received from thence; and yet this small difference was said to be the reason why the Rents of England fell one-fifth, that is 1,600 M. in 8 Millions. Which was a strange conceit, if they consider farther, that the value of the Cattel alive or dead, which went out of Ireland into England, was but 132,000, the Hides, Tallow, and Freight whereof were worth about ½ that money.

"2. Whereas the Owners of about ¼, both of all the real and personal Estate of Ireland, do live in England, since the business of the several Courts of Claims was finished in December, 1668, all that belongs to them goes out, but returns not.

"3. The gains of the Commissioners of that Court, and of the Farmers of the Revenue of Ireland, who live in England, have issued out of Ireland without returns.

"4. A considerable part of the Army of Ireland hath been sent into England, and yet paid out of Ireland.

"5. To remit so many great Sums out of Ireland into England, when all Trade between the said two Kingdoms is prohibited, must be very chargeable; for now the Goods which go out of Ireland, in order to furnish the said Sums in England, must for example go into the Barbadoes, and there be sold for Sugars, which brought into England, are sold for Money to pay there what Ireland owes. Which way being so long, tedious and hazardous, must necessarily so raise the exchange of Money, as we have seen 15 per cent. frequently given, Anno 1671 and Anno 1672. Altho in truth, exchange can never be naturally more than the Land and Water-carriage of Money between the two Kingdoms, and the ensurance of the same upon the way, if the Money be alike in both places."[1]

Proclamations were made from time to time forbidding the export of bullion and coin from Ireland,[2] but it is obvious that such proclamations could be of no effect in the circumstances.

In 1680 the value of foreign coin in circulation was fixed by proclamation as follows[3]:—

Coin.	Weight. Dwt.	Weight. Grs.	Value. £	Value. s.	Value. d.
GOLD.					
Rider	6	12	1	2	6
Spanish or French Quadruple Pistole	17	4	3	10	0
Double Ducat	4	12	0	18	0
Spanish Suffrain	7	2	1	8	6
SILVER.					
Ducatoon	20	16	0	6	0
Mexico, Sevil, or Pillar Piece of Eight, Rix Dollar, Cross Dollar, and French Lewis	17	0	0	4	8
Old Pern Piece of Eight	17	0	0	4	6
Portugal Royal	14	0	0	3	0

In the same year a new patent for making copper coins was granted for twenty-one years to Sir Thomas Armstrong and George Legge,[4] which the patentees transferred in 1685 to Sir John Knox.[5]

During the first three years of the reign of James II. no change was made in the system of coinage in Ireland. At the approach of the Revolution, a proclamation was issued raising the value of the foreign and English coins in circulation to the following amounts[6]:—

[1] *Political Anatomy*, pp. 71-2.   [2] Simon, pp. 52-3.   [3] Simon, p. 55.
[4] Simon, p. 54.   [5] Simon, p. 56.   [6] Simon, p. 57.

Coin.	Weight. Dwt. Grs.		Value. £ s. d.		
GOLD.					
Rider	6	12	1	4	0
French or Spanish Quadruple Pistole	17	4	3	16	0
Double Ducatoon	4	12	1	0	0
Spanish Suffrain	7	2	1	11	0
Guinea	—	—	1	4	0
SILVER.					
Ducatoon	20	16	0	6	3
Mexico, Sevil, or Pillar Piece of Eight, Rix Dollar, Cross Dollar, and French Lewis	17	0	0	5	0
Old Peru Piece of Eight	17	0	0	4	9
Portugal Royal	14	0	0	3	10
English Crown	—	—	0	5	5
English Shilling	—	—	0	1	1

The value of the English currency was raised by this proclamation in order to prevent the Protestant merchants from carrying it out of Ireland.[1] James next cancelled the patent which his predecessor had granted for the making of copper coins, and set up Royal Mints in Dublin and Limerick, where he issued a currency of mixed brass and copper shillings and sixpences, and, later, half-crowns of the same mixture. These debased coins were declared legal tender for all kinds of debts to any amount, and the possessors of good money were tempted to exchange it for the new bad money by the offer of a commission on the exchange.[2] A further issue followed of pennies and halfpennies made of a mixture of lead and tin, and of crown pieces of a white mixed metal. The net result of these issues was thoroughly to debase the whole coinage of the country.[3]

The debasement of the currency by James II. has been widely and loudly denounced by Williamite historians, and indeed, from the economic point of view, it was quite unjustifiable. But it is often forgotten that he was following the example of his illustrious predecessor Elizabeth, who, as we have seen, plunged the coinage of

---
[1] Simon, p. 57; Macaulay, *History of England*, vol. ii, p. 377, Everyman Edition.
[2] Simon, pp. 58-9.   [3] Simon, pp. 61-3.

Ireland into a state of complete discredit at the beginning of the same century. The only difference between the two cases was the motive. Elizabeth debased the currency deliberately, and from a distance, in order to impoverish the great majority of the Irish people by driving all their treasure abroad; James, on the other hand, debased it in the middle of a civil war, when he was in great personal danger, probably without any forethought of the effect of his action on the prosperity of the country, simply to provide himself with money to pay the army which was trying to save him his throne. It is admitted even by Macaulay that the measure was precipitated by the emigration of the richest class of the King's subjects at a moment when their presence and their wealth was most needed to save their country. In the one case the currency was debased in order to impoverish the rebels, whereas in the other it was debased because the rebels had themselves impoverished the Kingdom.

Banking still remained, as in the earlier part of the century, in the hands of brokers; it was very difficult to borrow money except at exorbitant interest. About the year 1680 credit suffered a shock owing to the failure of many of these bankers, and the erection of a public bank was seriously advocated. Although Ormond strongly advocated this project, it came to nothing.[1]

---

[1] Malcolm Dillon, *History of Banking in Ireland*; Lawrence, *The Interest of Ireland in its Trade and Wealth Stated*, vol. ii, p. 4; *Ormond MSS.*, vol. vii, p. 27.

CHAPTER IV.

THE PERIOD OF REDESTRUCTION, 1689-1700.

IRELAND was now destined to suffer the third of those great upheavals of her national life which took place within a single century, and which were such fatal bars to her orderly progress along the normal lines of economic development. We have seen that the last quarter of the sixteenth century was a period of absolute destruction of all forms of Irish wealth, during which the country was reduced to a condition of depopulation and poverty from which it gradually emerged during the first forty years of the seventeenth century. The great Rebellion then broke out, and during the next twenty years everything which had been built up in the beginning of the century was destroyed. After the Restoration the rebuilding process began again, and, as we have seen, the country developed rapidly and satisfactorily under Charles II. and James II. But its further development towards affluence was now to receive a rude check; in 1689 the Williamite war broke out; and in a few months the country was again denuded of thousands of its people and of the greater part of its wealth.

The loss which the population suffered during this outbreak was very serious, quite apart from the emigrations which followed. The course of the campaign has been told in many narratives which are familiar to all, and, although the various historians differ diametrically in their sympathies, and frequently in their accounts of what occurred, they are all agreed on the fact that the losses among King James's followers were very great. As usual, the war was accompanied by a plague,[1] and great numbers also perished from famine. A writer who observed the state of Ireland in the middle of the war was of opinion that, if things continued as they were going, two-thirds of the population would probably be wiped out by the sword, famine, plague, and emigration.[2]

This destruction of human life was accompanied by an equal, if not greater, destruction of property. One writer calculated that the losses of cattle incurred by the Protestants alone would amount to eight million pounds;[3] vast hordes of sheep were also destroyed. "The Irish tenants looked on the great sheep flocks with unfriendly eyes, and the first thing they did in the war was to destroy as many sheep as possible to make room for farms."[4] So great was the havoc wrought amongst the live stock of the country that Ireland, which for the previous twenty years had supplied Europe and America with provisions, was driven in 1692 to import cattle for its own use;[5] at the same time corn was so scarce as to be well nigh unprocurable.[6] The country was let run wild; tillage was rendered impossible by the fact that a great many horses had been seized for military purposes by the soldiers.[7] "There was destroyed in all parts of the kingdom above a million pounds of cattle besides corn and horses—twenty years of perfect peace could not restore the kingdom."[8] "English and French soldiers have already quite ruined this country, especially the latter by stealing and taking away of the country people's horses; the crops now rot in the ground for want of cattle to get them in. . . . . The peasants, where the soldiers are lodged, have hardly anything for themselves."[9]

[1] C.S.P., Dom., 1689-90, pp. 336, 368.
[2] *The Character of the Protestants of Ireland*, London, 1689.
[3] *The Character of the Protestants of Ireland*, London, 1697, p. 21.
[4] Arch. King to Mr. Annesley, *King MSS.*, 10th March, 1697.
[5] C S.P., Dom., 1691-2, p. 186. [6] *Leybourne-Popham MSS.*, p. 274.
[7] *An Exact Relation of the Persecutions, Robberies, and Losses Sustained by the Protestants of Killmare*, London, 1689; *Ireland's Lamentation*, London, 1689; *The Sad State and Condition of Ireland*, London, 1689.
[8] *Ireland's Lamentation*, London, 1689. [9] C.S.P., Dom., 1689-90, pp. 261, 531.

Towns and houses which had been rebuilt after the Restoration were razed to the ground. " Since the late troubles the condition of the country is much worse; many fair houses and some towns were burnt, and a great number of the people destroyed, so that of course manufactures must be impaired and lands untenanted."[1] The Lord Lieutenant, when opening Parliament in 1692, referred to "the almost utter desolation of the country caused by the Revolution."[2] "It will be impossible for the Irish to subsist if they have not speedy provisions from France, as the country is so harassed and the cattle driven away that the husbandmen have not enough cattle to plough the grounds."[3] In this general destruction of material wealth the woods suffered further damage.[4]

"The destruction of property which took place in a few weeks," wrote Macaulay, "would be incredible if it were not attested by witnesses unconnected with each other and attached to very different interests. All agreed in Dublin that it would take many years to repair the waste that had been wrought in a few weeks. Any estimate which can now be formed of the value of the property destroyed during this fearful conflict of races must be necessarily inexact. We are not, however, wholly without materials for such estimates. The Quakers were neither a very numerous nor a very opulent class; we can hardly suppose that they were more than one-fiftieth part of the Protestant population of Ireland, or that they possessed more than one-fiftieth part of the Protestant wealth of Ireland. Yet the Quakers computed their pecuniary losses at one hundred and fifty thousand pounds."[5]

The damage which the country suffered during the revolutionary war was, therefore, very great, but was as nothing compared with what it suffered during the peace, or rather the "war after war," which followed. We have seen that, after the Elizabethan and Cromwellian wars, in

---
[1] *The True Way to Render Ireland Happy and Secure*, Dublin, 1697.
[2] *I.C.J.*, vol. ii, p. 10.   [3] C.S.P., Dom., 1690-1, p. 120, and see pp. 152, 154, 155, 161, 191.
[4] Litton Falkiner, *Illustrations of Irish History*, p. 151.
[5] Macaulay, *History of England*, vol. iv, p. 165, 1858 Edition.

the course of which the country had been reduced to a state of extreme misery, things began to recover rapidly as soon as the war terminated; indeed, this was so marked a feature of Irish life as to attract the attention of many historians and to lead them to the conclusion that Ireland was possessed of some peculiar powers of recuperation. But if such power of recuperation still existed, it was given no chance of operating after the Williamite war, which was followed by a period of untiring persecution of the Catholics by the Protestants, and of Ireland by England. The great diminution of population which the country had suffered during the war was followed up by a still more serious diminution due to the emigrations after the war. After the siege of Limerick the greater part of King James's army went to France; the number of emigrants on this occasion was calculated to amount to fourteen thousand;[1] and this was but the beginning of a continuous stream of Catholic refugees who poured from Ireland to the Continent for many years. In 1692 we read that great numbers of Catholics were still emigrating,[2] and in 1694 the stream was still flowing.[3] An attempt was made to renew the Cromwellian system of selling the Irish as slaves, and shipping them to the colonies, and numbers of Irish boys and girls were transported in this way.[4] It was calculated that, between 1691, when the tide of emigration first commenced, and 1745, no less than four hundred and fifty thousand Irishmen had died in the service of France alone;[5] and great numbers were also to be found in the Spanish, Austrian, Neapolitan and other continental armies.[6] After 1699 another stream of emigration began as a result of the prohibition of the export of woollen goods in that year, when large numbers of Protestant artificers were driven abroad through lack of employment at home.

The lot of the Catholics who went abroad, however, was much less terrible than that of those who remained

---

[1] Lecky, *History of Ireland in the Eighteenth Century*, vol. i, p. 248.
[2] C.S.P., Dom., 1691-2, p. 67.   [3] *Buccleuch MSS.*, Montague House, vol. ii, p. 94.
[4] Moran, *Persecution of Irish Catholics*, p. 363.
[5] McGeoghegan, *Histoire d'Irlande*, vol. iii, p. 754.
[6] Lecky, *op. cit.*, vol. i, pp. 248-50; O'Callaghan, *History of Irish Brigades*.

at home, as the period was now beginning when the Catholic population of Ireland was to be ground down and degraded by the infamous system of penal laws, which have done more than anything else to injure the industrial character of the Irish people, and therefore the industrial wealth of Ireland. This orgy of religious intolerance was not only odious in itself, but was doubly odious, inasmuch as it was an express breach of the terms of the Treaty of Limerick, which provided that the Catholics of Ireland should be allowed the same religious liberty which they had enjoyed in the time of Charles II., and that they should be guaranteed against any disturbance on account of their religion.[1] The first step in the direction of the penal code was the British statute excluding Catholics from sitting in the Irish Parliament,[2] which thus transformed that body into an exclusively Protestant assembly. It is possible that the earlier measures of the penal code may have been dictated by a genuine fear of Papal domination, but, as measure after measure of totally unnecessary violence was enacted, it became apparent that the Irish Parliament was animated not so much by fear as by hatred, and that, as Burke thought, the code was based not on the insecurity of the Protestants, but on their security. "All the penal laws of that unparalleled code of oppression were manifestly the expression of national hatred and scorn towards a conquered people whom the victors delighted to trample upon and were not at all afraid to provoke. They were not the effect of their fears, but of their security."[3] "Intolerance," as Dr. Murray remarks, "largely prompted by political exigencies, becomes by indulgence an animating principle, much as infirmity of temper, beginning as a disease, passes into a vice."[4]

It is not necessary here to give any account of the penal laws themselves, as they have been frequently described in detail in other books, and as they more properly belong to the history of the eighteenth century.

[1] Lecky, *History of Ireland in the Eighteenth Century*, vol. i, p. 139.
[2] 3 Wm. and Mary, c. 2.   [3] Burke, *Letter to Sir Hercules Langrishe*.
[4] Murray, *Revolutionary Ireland and its Settlement*, p. 289.

Suffice it to say that they were such as to degrade the Catholics in every way, by taking away from them the property they had, and preventing them from acquiring more; by closing to them all the more estimable occupations and employments; and by depriving them of education. No Catholic was permitted to acquire any freehold interest in land or any leasehold interest of more than twenty-one years, and those who were already possessed of such interests were liable to be deprived of them at any moment by the defection of third parties. No Catholic could hold any civil or military office, nor could he practise any of the learned professions. In the sphere of trade, Catholics were discouraged by the regulations which prevailed in the corporate towns. In short, every road to prosperity was closed to them by an unparalleled code of laws which were designed " to make them poor and to keep them poor." Unfortunately, the code succeeded in attaining this object for over one hundred years.[1]

The penal code was primarily intended to deprive Catholics of the ownership of land, which was the chief wealth of the country, and we naturally pass from the laws against Catholics to a consideration of the changes which were wrought in the land system in the ten years following the Revolution. It may be objected that no mention has been made so far in this book of the land legislation attempted by the Patriot Parliament, but the reason that this has been omitted is that, whatever its significance may have been from a political point of view, it is of no importance to the economist, for the simple reason that it was never carried into effect. Much abuse has been showered on King James's Parliament for its attempted dealings with the land, and, indeed, nobody will deny that it was a foolish and ill-advised proceeding to attempt in the midst of a civil war to revolutionise the existing land ownership in the country. But it must be remembered that the members of that Parliament had been brought up

[1] O'Brien, *Economic History of Ireland in the Eighteenth Century*, pp. 24 to 30.

in a school where scant attention was paid to rights accruing from long possession. They were the sons of the men, frequently the very men themselves, who had been violently dispossessed by Cromwell, and cheated of their right of restoration under the Act of Settlement, and they can hardly be blamed for attempting to apply to their opponents the same principles that had been applied against themselves. Moreover, it must also be remembered that the Act repealing the Act of Settlement was in two respects more enlightened and just than previous dealings with Irish land had been. In the first place, it made provision for the compensation of those who in the interval between the Acts of Settlement and the Patriot Parliament had acquired purchasers' interests in the land; and in the second place it expressly preserved the rights of all tenants who had taken leases of the confiscated lands.[1]

The Williamites after their victory showed themselves no more scrupulous in dealing with vested interests in landed property than the Jacobites had been. Enormous tracts of Catholic property were confiscated as a result of the revolutionary war; there is no necessity here to examine the exact amount of land which changed hands at this period, about which there is some doubt; it is sufficient to say that in the year 1700 the total quantity of Irish land which still remained in the hands of Catholic owners was not more than one million English acres.[2] That this quantity could never be increased, but would almost certainly, in course of time, be diminished, was rendered certain by the penal laws. Any Protestant lady being heiress to or in possession of any real property who married a Catholic lost all her lands;[3] any Catholic whose eldest son became a Protestant became but a tenant for life in his estates; and a Catholic's land, descending by intestacy, descended not to his eldest son but to all his sons equally.[4] These laws tended to diminish the

[1] Davis, *Patriot Parliament*, pp. 88, 103-4, 107.
[2] Butler, *Confiscations in Irish History*, p. 232.
[3] 9 William III, c. 3.   [4] 2 Anne, c. 6.

quantity of land in Catholic hands, while, by other statutes, it was also provided that that quantity should at no time be increased.¹ When the confiscated lands were put up for auction in 1701 the Irish were prohibited from becoming purchasers of more than two acres even as tenants.²

Simultaneously with the confiscations of the Catholics' lands the rights of tenants were attacked. We have seen that, in the course of the seventeenth century, various customary rights between landlord and tenant had been recognised, and that the tenants felt so secure in their reliance on these customs that the taking of long leases was rather the exception than the rule. It is probably correct to say that the majority of the tenants in Ireland held their lands, not under any lease, but relying on " the custom of the country." In 1695 was passed what Dr. Sigerson describes as "a remarkable degrading Act, which precipitated immense numbers of formerly estated tenants into the dependent and serf-like condition of tenants at will."³ By this Act, all leases and other interests in land made by livery of seisin or by parole and not put in writing were declared to be merely tenancies at will, with the exception of leases not exceeding three years, of which the rent should be two-thirds of the value of the thing demised, or dealings in interests of copyhold or customary tenure.⁴ The effect of this Act was to put the old protected tenants of the country at the mercy of their new landlords. The exception in favour of short leases at high rents is significant, as it shows that the objection was not to unwritten leases but to low rents. This Act undoubtedly was the first step towards the ultimate abrogation of tenant right, and furnished a convenient weapon to grasping and hostile landlords and their agents. It had the effect of turning the greater part of the tenantry of the country at one blow into the position of mere tenants at will, and it is significant that shortly after the passing of the Act it was necessary to pass another Act dealing

---

¹ 2 Anne, c. 6, sec. 6; 8 Anne, c. 3.   ² Prendergast, p. 24.
³ Sigerson, *Land Tenures in Ireland*, p. 104.   ⁴ 7 William III, c. 12.

with the suppression of agrarian outrages.¹ A similar Act had been passed in England in 1677 with the avowed intention of extinguishing numerous small holders at customary rents, and this Act did much to abolish the yeoman farmers of England.² If the Act was felt as a hardship in England, how much more of an evil must it have been in Ireland where the landlord class looked down upon and hated their Catholic tenantry, and where the law was administered by partial and bigoted tribunals?

The fatal ten years with which we are now dealing, which witnessed the emigration of so many thousands of Irishmen, and the dispossession and degradation of so many more who remained at home, were also marked by the beginning of the systematic war waged by England on Irish industrial prosperity. In Ireland itself there was a genuine desire on the part of the Government to increase the country's industry and trade. The Committee of Trade was still in being, and was so occupied with projects and suggestions that sub-committees had to be appointed to consider the various matters submitted to it.³ William III. was also anxious to advance Irish trade, and sent instructions that the fishing, linen, and provision trades should be encouraged as far as possible.⁴ No discouragement was intimated in any speech from the throne in the Irish Parliament during these years. "Their Majesties," said Lord Sidney in 1692, "being in their own Royal judgment satisfied that a country, so highly favoured by nature, and so advantageously situated for trade and navigation, would want nothing but the blessing of peace and the help of some good laws to make it as rich and flourishing as most of its neighbours; I am ordered to assure you that nothing shall be wanting on their parts that may contribute to your perfect and lasting happiness."⁵

The attack on Irish trade, however, did not come from any party in Ireland; on the contrary, all the inhabitants

---
¹ 7 William III, c. 21.
² Gibbons, *Industrial History of England*, London, 1892, p. 114.
³ *I.C J.*, vol. ii. pp. 32, 79.   ⁴ C.S.P., Dom., 1691-2, p. 169.
⁵ Hely Hutchinson, *Commercial Restraints*, p. 56.

of the country were naturally anxious to increase its prosperity. The attack on Irish trade was an attack directed from England. We have seen, when dealing with the land and the penal laws, that considerable energy was devoted by the Irish Protestants to impoverish the Irish Catholics, but now the oppressors were themselves to be the oppressed. The cat had enjoyed the bird, but it was now the dog's turn to relish the cat. It must be constantly borne in mind when considering this question that the attacks made by the English Parliament on Irish trade were attacks made, not on the Catholics, but on the Protestants of Ireland, who, by their own oppression of the Catholics, had succeeded in securing for themselves most of the industry and trade of the country. It was on this account that the Protestant interest in Ireland so bitterly resented these attacks; and, indeed, it was this commercial oppression which first gave rise to the Irish Protestant nationalism which afterwards became so strong and so patriotic in the later eighteenth century. These attacks must have appeared to the Irish Protestant as something in the nature of a stab in the back. They who had been at pains to subdue Ireland, as they thought, in the Protestant interest, now found themselves in process of being subdued in the English interest. "The measures against Irish trade," wrote Archbishop King, "will be particularly ruinous to the Protestant English interest of Ireland; inasmuch that they tend to alienate the affections of the King's subjects from his Majesty's and discouraged them from their vigorous prosecution of popery whereby Ireland might be effectually secured to England without danger of rebellion. . . . The principal losers will be the English gentlemen and tradesmen."[1] That the injury inflicted by the trade restrictions was principally confined to the Protestant interest in Ireland is also apparent from every page of the political writings of Swift and Hely Hutchinson, neither of whom had any desire to benefit any of the inhabitants of Ireland except Protestants; and

[1] *King MSS.*, 2nd April, 1698.

from the course of the free trade agitation of 1779, the principal actors in which were bitterly opposed to any measures which would improve the position of the Catholics.

The attack on the woollen industry was particularly calculated to injure the Protestants, as this industry was almost altogether in their hands. In 1695 it was said that the manufacture of wool was "not spread amongst the Catholics of the country at all,"[1] and in the same year it was said that there were twelve thousand English families in Dublin and fifty thousand in Ireland engaged on the woollen manufactory.[2] Indeed, the Protestants bitterly resented the intrusion of any Catholics into this industry, as appears from a petition which they presented to the Irish House of Commons.[3] The Catholics, for their part, were not ill pleased at the check to the woollen manufacture; they regarded the sheep as occupying their place upon the land. This explains why, in the Revolution of 1689, they destroyed huge flocks of sheep. One of the results of the destruction of the woollen industry was that Catholics began to be reinstated as tenants on lands, which had theretofore been devoted to sheep pasture.[4] The Protestant woollen workers, on the other hand, were driven from their employment and forced to emigrate in large numbers.[5] The folly of this legislation was clearly seen in Ireland at the time:—"Make Ireland rich and they will be your bees and bring honey to your hives; keep them poor and that kingdom will provide nothing but wasps to sting you; the English will leave the country, and then it will fall into the hands of the Irish."[6]

The explanation of this commercial policy is to be found in the victory of the Parliament over the King in the Revolution. Under the Stuarts it had been the constant policy of the English sovereign to raise a large revenue in Ireland with which to render him less dependent on

---

[1] *A Discourse on the Woollen Manufactory of Ireland,* Dublin, 1698; *A Discourse Concerning Ireland,* London, 1698.
[2] *A Discourse Concerning Ireland,* 1697-8.   [3] *I.C.J.,* vol. ii, p. 247.
[4] Sigerson, *Land Tenures in Ireland,* p. 97.   [5] Lecky, *op. cit.,* vol. i. p. 440.
[6] *Anonymous Paper in Buccleuch MSS.,* Montague House, vol. ii, p. 745.

Parliamentary support in England, and on two occasions Irish troops, paid for out of Irish taxes, had actually been brought to England to fight against the popular party. The Parliament of William III. was determined that no such power should ever be exercised by the King again, lest Ireland should become a danger to England. " In the days of Charles I., troops from Ireland had landed in order to aid an absolute monarch. Irish troops had also pitched their camp on Hounslow Heath to aid an absolute monarch's despotic son . . . . the English realised that they must avert such risk, and the plain way seemed to be systematic depression of industries in that land if her interest in any way conflicted with the interests of the country."¹ The importance of this consideration is emphasised by Dr. Cunningham.² " Unfortunately, the economic jealousy with which Englishmen regarded Irish progress was universally stimulated by considerations of a constitutional character. The English Party was keenly suspicious of anything that might tend to increase the royal powers. Charles I., Charles II., and James II. had all suffered from the distrust of their subjects; and William III., even though he had been invited to come over, did not succeed in inspiring confidence. As is well known, he bitterly resented the treatment he received. Since Ireland was an independent kingdom, the English House of Commons had no direct control over its affairs; and there was constant uneasiness lest any power which the King acquired in Ireland should be used without the concurrence of the English Parliament, or even against English liberties. Twice within the seventeenth century serious attempts had been made to develop the resources of Ireland—by Strafford, and under Charles II. and James II.; in both cases the result had been that the King had found himself in possession of power that seemed to menace his English subjects. Under these circumstances there was the strongest poli-

¹ Murray, *Revolutionary Ireland and its Settlement*, p. 307; Murray, *Commercial Relations*, p. 51.
² *Growth of English Industry and Commerce*, vol. ii, p. 371.

tical reason for dreading any development of the wealth of Ireland that took place at the expense of England, since this really implied an increase of the influence of the Crown at the expense of that of Parliament. Traces of this feeling were found in 1779 and later, as, for example, in speeches by Lord Shelbourne,[1] and by Fox."[2] The same policy was dictated by the further motive of fear and hatred of France, which had been the political ally of Ireland in the Revolution, and which might be in a position to aid Ireland materially at some future date, if Ireland were rich enough to pay for such assistance.[3]

These were the two most powerful motives which induced the English Parliament to adopt a policy of systematic hostility to Irish interests in trade, but they were strengthened by trade jealousy, and the fear that Irish prosperity might injure England. Such jealousy was common at a period dominated by mercantile ideas, but additional force was given to it at the moment because in the opinion of the English Parliament English commercial prosperity must be maintained at all costs in order to enrich England in its great struggle with France which was then proceeding :—" England was waging a vital struggle with a wealthy opponent, and, if her commercial resources were impaired, her chance of ultimate success were to that extent destroyed."[4]

In order to carry these hostile intentions into effect, it was necessary that the political status of Ireland should be reduced from that of a kingdom to that of a colony. The essence of successful mercantilism was that the country struck against should not be in a position to strike back, and, therefore, the power of the English Parliament to legislate for Ireland was insisted on during this period, and the independence of the Irish Parliament was undermined. Whatever had been the relation of Ireland to England before the Revolution, its position was now clearly defined as that of a colony, the

[1] *Parl. Hist.*, vol. xx, p. 1163.   [2] *Ibid.*, vol. xxi, p. 1297.
[3] Murray, *Revolutionary Ireland and its Settlement*, p. 385.
[4] Murray, *Revolutionary Ireland and its Settlement*, p. 390.

interests of which in all things commercial must be subordinate to those of the mother country. Ireland, therefore, took its place in the colonial fabric of the British Empire, and its interests were dealt with by the English Parliament in exactly the same way as were the interests of America. This was the age of the growth of the Imperial conception, by which it was hoped that the productive power of the Empire as a whole should be utilised with the greatest possible effect. " In theory one form of production would be assigned to America, another to Ireland, another to England. In practice, England took what was convenient or agreeable to herself; Ireland and the colonies had the leavings—in Ireland there was worse abuse than restriction—the destruction in the interests of England of existing industries."[1]

The degradation of Ireland to the status of a colony had been foreshadowed in the Navigation Acts, and it was now fully and completely carried into effect. The first duty of a colony in those days was to produce raw material for the manufactures of the mother-country. This principle was applied to Ireland in 1695, when the duties in England on bar iron unwrought imported from Ireland were taken off,[2] and also by the whole course of dealing with the export of Irish raw unmanufactured wool, which, as we have seen, was designed to reach no destination but England, whose woollen manufacture it was designed to feed. Some years after the Revolution the colonial theory was laid down in full form. "An inscription of the purpose following should be always set up in the Irish House of Commons to be read the first thing every day of the session: 'Let us always remember that this island is a colony; that England is our mother country; that we are ever to expect protection from her in the possession of our lands, which we are to cultivate and improve for our own subsistence and advantage, but not to trade to or with any other nation without

[1] Meredith, *Economic History of England*, p. 191.
[2] Murray, *Commercial Restraints*, p. 85.

her permission; and that it is our incumbent duty to pay obedience to all such laws as she shall enact concerning us.' "[1]

The application of the colonial theory by England to Ireland involved peculiar injustice, because the two countries had reached almost the same state of industrial development, and both produced the same commodities for export. It was one thing to prohibit the exportation of worked-up commodities from America, where practically no manufactures existed; another to prohibit their exportation from Ireland, where industrial activity was beginning to develop.

It is not surprising that this campaign against Irish industry was preceded and accompanied by a campaign directed against the legislative independence of the Irish Parliament, as the former would have been unsuccessful but for the latter. \ It was in the years immediately following the Revolution that the right of the English Parliament to legislate for Ireland was for the first time strenuously advocated and definitely insisted on,\in spite of the strong opposition of Molyneux and other Irish writers, who saw the significance of this new development and guessed in what direction it was leading.[2]

The first blow in this campaign against Irish industrial prosperity was, as might be expected, aimed at the woollen industry, which was the most considerable industry of that date. Although this industry had been encouraged and, indeed, for all practical purposes, brought into being by the action of England in prohibiting the importation of Irish cattle, it had always been a subject of considerable jealousy amongst those who had themselves produced it. As early as 1676, petitions were presented in England for the forbidding of all Irish manufacture;[3] and in the following year an Irish writer prophesied that, as the import of Irish cattle to England had been looked on as a nuisance, and the export of raw

[1] *The Interest of England as it Stands with Relation to the Trade of Ireland Considered*, London, 1698.
[2] Swift McNeill, *History of the Irish Constitution*, p. 10.
[3] C.S.P., Dom., 1676-7, p. 386.

wool from Ireland as a felony, in a few years the erection of a woollen manufactory would be regarded in England as nothing less than treason.[1]

But, however jealous the English woollen merchants might be of the growing Irish trade, they were unable to suppress it before the Revolution. After that event their jealousy became much more marked. In 1691 it was urged that it would be a wise measure to exchange the woollen manufacture for the linen in Ireland " to furnish against even the remotest possibility of detriment ";[2] and English jealousy was given full voice in a pamphlet which appeared four years later:—" That Ireland is now destructive to the interest of England I think will admit of little dispute; for so long as that people enjoy so free and open a trade to foreign parts, and thereby are encouraged to advance in their woollen manufacture, this must consequently lessen ours, than which they cannot do us a greater mischief, being the tools whereon we trade. When they sink our navigation sinks with them. Now the advantage Ireland hath above England in making the woollen manufacture will soon give them opportunity of outdoing us therein, first as it produces as good or rather better wool, and next as it furnishes all provisions cheaper to the workmen, which renders them able to live on easier terms than ours can here, and this will, in short time, give invitation for many more to remove thither."[3] " With this end in view, the act of prohibiting cattle should be repealed."[4] The campaign was waged with more violence than logic; any argument was good enough to use against Ireland, even if it contradicted the writer's own previous argument. For instance, one of the reasons advanced by Cary for the suppression of the woollen industry was that it was a benefit only to the trading class, and not to the landowners, who were the real supporters of the English interest; but a few pages

---

[1] *Letter from a Gentleman in Ireland to his Brother in England Relating to Trade*, London, 1677.
[2] *Remarks on the Trade and Interest of England in Ireland*, London, 1691.
[3] Cary, *An Essay on the State of England in Relation to its Trade*, pp. 91-2 Bristol, 1695.   [4] *Ibid.*, p. 100.

later he urged that the Cattle Acts should be repealed, in order to benefit the English woollen industry, which would be productive of the greatest gains for the landowners of England. Thus the industry should be destroyed in Ireland because it was of no benefit to the landlords, and encouraged in England because its growth would benefit them so greatly.

No doubt, the flame of this jealousy was fanned by the increasing prosperity of the Irish woollen industry after the Revolution, owing to the fact that many of the Irish Protestants who had emigrated during the Jacobite wars had learned the trade in the West of England, and, when the war was over, had brought their knowledge back to Ireland.[1] The attention paid to the interests of the woollen industry in the Irish House of Commons shows that it must have been of considerable dimensions and worth much consideration.[2] The quality of Irish woollens must have been acquiring a reputation abroad; in 1702 hangings of Irish woollen stuffs were used in the palace at Copenhagen.[3] In 1698 it was said that there were twelve thousand Protestant families in Dublin, and thirty thousand in the rest of Ireland engaged in the woollen trade,[4] and, while this is probably an exaggeration, as Dr. Murray has pointed out, it shows that the industry was certainly widespread.[5] The following table[6] shows the exports of woollen goods from 1690 to 1698:—

EXPORTS.

Year.	Drapery, New Pieces.	Drapery, Old Pieces.	Frieze.	Stockings, Dozens.	Wool Yarn	Wool.	Rugs.	Hats.
1690	247	11	101,419	820	—	—	3	—
1691	1,470	50	150,691	1,641	—	—	5	—
1692	1,500	62	62,771	1,618	—	—	—	—
1693	2,726	23	34,681	898	1897	36,888	53	—
1694	2,912	28	20,839	2,370	1492	38,794	—	—
1695	2,608	17	41,146	1,251	883	69,751	—	—
1696	4,413	34¾	104,167	2,919	7,900	89 783	144	—
1698	23,285½	281¼	666,901	770	—	—	458	4,470

[1] E.C.J., vol. xii, p. 63.   [2] I.C.J., vol. ii, pp. 725, 733; vol. iii, pp. 45, 65.
[3] Portland MSS., vol. ii, p. 59.   [4] O'Conor, History of Irish Catholics, p. 149.
[5] Murray, Revolutionary Ireland and its Settlement, p. 394.
[6] Murray, Revolutionary Ireland and its Settlement, pp. 395-6.

IMPORTS.

Year.	Drapery, New Pieces.	Drapery, Old Pieces.	Stockings.	Value of New Drapery, £	Value of Old Drapery, £
1693	90,259	14,504	4,710		
1694	49,620	13,085	—		
1695	672,932	136,562	1,098		
1696	45,064	15,227	2,874	2,043	9,014
1700	24,522	12,119			
1706	15,308	5,514		1,913	4,135

It is obvious from these tables that, though the woollen manufacture was growing, Ireland was still importing more woollens than she was exporting, and could not have been a serious danger to the English industry. Indeed, it is quite clear from the petitions of the English wool workers that the danger apprehended was more in the future than in the present.[1] " The apprehensions of England seem rather to have arisen from the fears of future, than from the experience of any past rivalship in this trade."[2]

Nevertheless, in spite of the remoteness of the danger apprehended, the alarm in England was great. Numerous petitions were presented to Parliament by the woollen manufacturers complaining that many English weavers were emigrating to Ireland where goods could be produced much cheaper than in England, and that the result of this would undoubtedly be that English woollens would be undersold in the foreign market.[3] The House of Commons itself petitioned the King in the same sense: —" We cannot without trouble observe that Ireland, which is dependent on and protected by England in enjoyment of all they have, should of late apply itself to the woollen manufacture to the great prejudice of the trade of this Kingdom . . . . we humbly implore that you will make it your royal pleasure for the discouraging

[1] E.C.J., vol. xii, pp. 37, 63, 64.   [2] Hely Hutchinson, *Commercial Restraints*, p. 60.
[3] E.C.J., vol. xii, pp. 37, 40, 63, 64.

the woollen manufacture and encouraging the linen manufacture in Ireland."¹ The King promised in his turn to do all that he could to injure the Irish woollen industry, and a message to this effect was conveyed to the Irish House of Commons by the Lords Justices at the beginning of their new session. As a result, an Act was passed in Ireland imposing for three years an additional duty of four shillings per pound on broadcloth exported and two shillings per pound on new drapery made or mixed with wool.² This Act, however, was not sufficient to satisfy English jealousy, and in the following year an Act was passed in the English Parliament prohibiting perpetually the exportation from Ireland of all goods made or mixed with wool except to England.³ The exception in favour of exportation to England was really a nullity, as the heavy protective duties imposed in 1660 were still retained. The results of this measure were disastrous in the extreme, but, as they did not show themselves until the eighteenth century, we are not concerned with them here.

It has been frequently argued that the conduct of England in suppressing the Irish woollen industry should be excused on the ground that the linen industry was established instead. To this suggestion there are two answers, first, that an unjust deprivation is not rendered any less unjust because some compensation is offered for the thing taken away, and secondly that the consideration in this case was not in any sense adequate. The first of these propositions requires no elaboration. The English Parliament had no constitutional or legal right to impose any law forbidding the exportation of goods from a country not under its jurisdiction; and the illegality of its proceedings on this occasion was rendered doubly unjustifiable, inasmuch as they were dictated by selfish and jealous motives. It had no more right to direct Irish capital into the linen industry than it had to divert it from the woollen. Indeed, the essence of the injustice was the

[1] B.C.J., vol. xii, p. 338.   [2] 10 William III, c. 5.   [3] 10 & 11 William III, c. 10.

interference with the livelihood of a people by a body in which that people was not only not represented, but was regarded with hostility and contempt. It may or may not be desirable that my neighbour should possess a gun with which he may possibly injure me in the distant future, and in a combination of circumstances which has not yet arisen; but it is quite plain that I am not entitled to deprive him of that gun by main force and give him a feather in its place.

In any event the encouragement of the linen industry was not an adequate compensation for the destruction of the woollen industry. The latter was a manufacture peculiarly suitable for Ireland; raw material was produced in the country in great quantities and of excellent quality; and the foundation of its success had been laid by the labour and enterprise of many years. The linen industry, on the other hand, was concerned in the working up of a material which was never produced in Ireland as successfully as elsewhere; it had attained to whatever position it held at the time by reason of much artificial encouragement and support; and, as we now know, it ultimately failed to spread to more than a small part of Ireland. The exchange was one of a certainty for an experiment.

It must also be remembered that the successes of the two industries were not mutually exclusive; there was no reason why they should not have both advanced together. Indeed, there is little doubt but that they would have done so, had they been allowed. The period of Grattan's Parliament was marked by an extraordinary revival of the woollen manufacture, and the same period also saw an unprecedented expansion of the linen manufacture.[1] If the two industries were capable of progressing side by side in 1782, there was no reason why they should not have done so in 1699.

To promise to give Ireland a linen industry was really to promise her something which she already had. The

[1] O'Brien, *Economic History of Ireland in the Eighteenth Century*, pp. 269-75.

fact that her linen industry was smaller and of less importance than her woollen is beside the point, as the two had probably grown in dimensions proportionate to their respective suitability.

We have seen the progress which the linen manufacture made under Charles II., and this progress was maintained after the Revolution. It is true that the export of Irish linen did not attain any considerable dimensions; the total exports in 1700 only amounted in value to £14,112.[1] It is also true that a venture to increase the manufacture had failed in 1698; but this failure was caused by the stock-jobbing and insufficiency of capital of the English company which had control of the enterprise.[2] As against this, it must be remembered that the growth of the industry in Ireland attracted the attention of all contemporary observers; in 1691 it was said to be one of the most considerable of the manufactures in the country;[3] and it sprang up simultaneously at Dublin, Drogheda, and in many parts of Ulster.[4] The Scotch settlers made great quantities of linen cloth, which they exported to England: "The commonalty of them are so intent upon this kind of manufacture that the very husbandmen and their servants, when they return from their labours abroad, do employ themselves by their firesides in this kind of work, and sit reeling of linen yarn while their women are busy spinning; and by their constancy and diligence that country produces great quantities of good linen yearly."[5] In 1695 all kinds of flax and hemp were permitted to be imported from England, duty free,[6] and we may deduce from this permission that there was a considerable demand for these materials in Ireland. In the following year, Molyneux wrote that looms and bleaching yards were being widely established, and that much fine

---

[1] *I.C.J.*, vol. xvi. p. 362.
[2] A full account of the career of this Company will be found in the *Kilkenny Archæological Journal*, 5th Series, vol. xi, p. 371; see also Murray, *Revolutionary Ireland and its Settlement*, p. 4(6.
[3] *Remarks on Interest and Trade of England and Ireland*, London, 1691.
[4] *Kilkenny Arch. Jnl.*, 5th Series, vol. xi, p. 371; *A Discourse Concerning Ireland*, London, 1697-8.
[5] *A Discourse Concerning Ireland*, London, 1697-8.   [6] 7 & 8 Wm. III., c. 39. (Eng)

linen was being produced. "I have as good diaper made by some of my tenants near Armagh as can come to a table, and all other cloth for household use."¹ Linen head-dresses were used by the majority of Irish women.² Finally, it must be remembered that Crommelin and his followers had settled down at Lisburn some years before the English Parliament promised to give Ireland the linen manufacture in exchange for the woollen.

It is a matter of some difficulty to form an opinion as to the dimensions of the linen industry in Ireland in the years immediately following the Revolution. The statements on the subject are contradictory, and reliable statistics are not to be obtained. However, whether the industry was large or small is a matter of no importance in considering the justice of English policy towards Irish trade. If the industry was small, then Ireland was being forced to exchange a certainty for an experiment; if it was large, then the pretended consideration for the destruction of the woollen industry was no consideration at all.

In the general convulsion of Irish affairs which accompanied and followed the Revolution, the Irish revenue also suffered. In the period from 5th June, 1690, to 29th September, 1692, the revenue was £277,217, and the expenditure £879,966.³ The succeeding years also showed deficits, which were made up partly by remittances from England, partly by borrowing, and partly by increased taxation. The remittances from England were not continued longer than was absolutely necessary for the military safety of Ireland, and, as soon as it was possible to do so, the Irish people was made to bear the expenses of the augmented army which was being maintained. In the year 1692 the Government raised £33,050 by a loan at ten per cent. secured on the quit rents.⁴

The bulk of the additional expenses were met by the imposition of increased taxation—the "additional duties.' The first of these were imposed in 1692, when an addi-

---
¹ *Locke's Works*, vol. iii, p. 552. ² *A Brief Character of Ireland*, London, 1692.
³ Official Accounts of Receipts and Expenditure, *Parliamentary Papers*, 1868-9, vol. 35. ⁴ *Ibidem*.

tional excise duty of 1s. 6d. a barrel was put on beer worth more than 6s. a barrel, threepence on beer of less value, and threepence a gallon on aqua vitæ and strong waters.[1] The increase of revenue produced by these duties did not balance the increased expenditure, and it was consequently necessary to impose further taxes in 1695. On this occasion resort was had to a poll tax, which provided that a shilling should be paid by every individual in the country, with the exception of the wives and daughters of day labourers living with their parents, labourers' sons under eighteen, widows excused from paying hearth money, and those living on alms. In addition to this universal tax, all persons of station or possessed of property had to pay a further tax at a graduated rate. It is interesting to note that the principle of the taxation of bachelors was recognized by this Act, which provided that a double tax should be paid by traders who were not freemen, and by bachelors over thirty.[2] In the following session this poll tax was continued, but the amount payable was doubled.[3] This tax was calculated to bring in a large revenue, but it must have been very carelessly collected, as the total amounts produced were very inconsiderable[4]:—

Year Ending 25th Dec.	£
1696	7,678
1697	9,961
1698	5,615
1699	5,666
1700	3,689

In spite of all this additional taxation, the revenue still proved unequal to the expenses of the Government, and in 1698 the experiment of a land tax was tried.[5] £120,000 was to be raised by four half-yearly payments of £30,000; to each of these payments Leinster was to contribute £10,050, Munster £8,940, Ulster £7,000, and

[1] William and Mary, c. 3. [2] 7 William III, c. 15. [3] 9 William III, c. 8.
[4] *Parliamentary Papers*, 1868-9, vol. 35. [5] 10 William III, c. 3.

Connacht £4,010; and each barony was to bear the proportion at which it has been assessed by the presentments of Grand Juries at Assizes and Quarter Sessions.

The only other additional duties which remain to be mentioned are a small additional excise on tobacco to defray the cost of building and repairing military barracks,[1] and the additional duty to which we have already referred on the export of woollen goods.[2] This last duty was not imposed for the purpose of increasing the revenue, and any chance which it had of doing so was prevented by the prohibition in the following year of the exportation of woollens from Ireland.

Thus, the period following the Revolution was characterized by increased taxation, together with decreasing wealth, a thoroughly unsatisfactory condition, unfortunately not unfamiliar to those acquainted with Ireland. "On a review of the Parliamentary events which occurred in the reign of William III., particularly in the department of finance, there will be found but few periods in the annals of our country which appear with so much disadvantage."[3]

*　　*　　*　　*　　*　　*

We have now completed our review of the economic condition of Ireland in the seventeenth century. In the ten years following the Revolution, which form the subject of the present chapter, was sown the terrible crop which was reaped in the eighteenth century. The event which may be taken as marking the close of the period we have studied was the suppression of the Irish woollen manufacture, which in many ways was the most important landmark in the whole economic history of Ireland. It certainly did more to shape the course of Irish economic life in succeeding years than any other single event, and was the

---

[1] 10 William III, c. 4.   [2] 10 William III, c. 5.
[3] Clarendon, *Revenue of Ireland*, 1791, pp. 34-5.

most fruitful source of the dreadful distress that characterised the eighteenth century.

It is necessary, however, to qualify this statement in one respect, as too much emphasis has been laid by some writers on the suppression of the woollen industry in itself. The fact is that the prohibition of the export of Irish woollens was attended with such disastrous consequences, not so much because of the prohibition itself, as because of the condition of Ireland in other respects. The suppression of the woollen industry was attended with disastrous consequences, because it practically destroyed all manufacturing industry in Ireland, and therefore threw the whole population on the land for subsistence. It is conceivable that, had the land system of Ireland been based on a foundation of equity, and regulated by just laws, the whole population of Ireland might have succeeded in deriving a comfortable livelihood from the pursuit of agriculture alone; but, as it was, the many evils which characterized the Irish land system, chiefly the prevalence of absenteeism, the universal existence of rack rents, the encouragement of pasture at the expense of tillage, and the penal laws, operated to produce a state of affairs in which the land afforded only a bare subsistence to the vast majority of those engaged in its cultivation. It was, therefore, by placing the whole population in a position in which it was exposed to the evils of the land system that the suppression of the woollen industry produced its disastrous effects. It may be remembered that Bishop Berkeley suggested in the *Querist* that the importance of the suppression of the woollen industry was generally exaggerated by the economists of his day, but he did not indicate, possibly because he failed to see, the means which the landowners of Ireland could have taken to minimise the ill-effects which that measure had undoubtedly produced.

Subject to this qualification, however, and bearing in mind the importance of understanding that the suppression of the woollen industry derived its chief importance from the circumstances of Ireland in other respects, it may be safely stated that the suppression of that industry was the most important event in Irish economic history. It certainly formed the dividing line between the seventeenth and eighteenth centuries. It terminated the era of hope, and inaugurated the era of despair.

The seventeenth century in Ireland was characterized by periods of economic progress nullified by political cataclysms. The century opened on a spectacle of ruin, resulting from the prolonged and devastating wars of Elizabeth's reign; for forty years a determined and, on the whole, successful effort was made to evolve order out of chaos; but the progress of forty years was undone by the ravages of the Rebellion years. Once more Ireland passed through a period, of which the economic history may be summarised by saying that those who escaped death from war, famine, and disease, succeeded in sustaining life at the bare margin of subsistence by the fitful and often interrupted cultivation of the soil. After the Restoration Ireland was blessed with thirty years of peace, during which extraordinary progress was made, in spite of discouragement in many directions. It looked as though a stable and prosperous country were going to be built up at last, but all the efforts of a generation were again frustrated by another political upheaval. This outbreak, though itself not so destructive to life and property as the previous wars had been, was attended ultimately by far graver consequences, as the peace which followed was not peace, but rather the " war after war," of which we hear so much at the present day. Thus, three times in the course of a single century, the orderly and normal economic progress of Ireland was interrupted by political cataclysms.

The extraordinary recuperative power which Ireland displayed after the Elizabethan and Cromwellian wars attracted much attention at the time, and has been commented on by many historians. No such recuperative power, however, showed itself in the period following the revolutionary war, and its absence shows more strikingly than anything else could have shown the profound change for the worse that had come in the condition of Ireland between the beginning and the end of the seventeenth century. It cannot be suggested that the character of the people had changed, or that the natural resources of Ireland had diminished, and we must therefore conclude that the change had its origin in some external circumstance. The fact is that a profound change had taken place in the Irish policy of the English Government. The Stuart Kings were always anxious to develop the resources and to increase the wealth, and consequently the revenue of Ireland, with a view to rendering themselves less dependent for financial support on the English Parliament. With the overthrow of the Stuarts, Parliament was supreme in England, and was quite determined that it would not risk any assault on its prosperity by allowing the King to obtain a large revenue from any source outside its control, as for instance he might have done from Ireland. This point has been made so frequently in the above pages that it is not necessary to do more than to mention it here. It is true that the schemes for the betterment of Ireland under the governments of the Stuarts involved many profound injustices; that the prosperity that it was hoped Ireland would enjoy was meant to accrue for the benefit of the English settlers; and that the rightful owners of the soil were dispossessed of their land and degraded to the position of tenants, or driven to emigration; but it nevertheless remains equally true that the policy of the rulers of Ireland at the beginning of the seventeenth century was to turn the country into a garden, while the policy of its rulers at the end of the same century was to turn it into a wilderness.

In conclusion, it must ever be remembered that the realization of the cruel ambitions of the statesmen who succeeded the Revolution was only rendered possible by the destruction of the independence of the Irish legislature, and that the era of trade restriction and economic repression was heralded by a successful, if unconstitutional, assertion of the right of the English Parliament to legislate for Ireland. The more one studies Irish history, especially Irish economic history, the more one becomes convinced of the profound truth of Isaac Butt's observation :—" Never perhaps was the physical misery of a country more directly connected by clear and overwhelming evidence with its national destruction and its political degradation."[1]

---

[1] *The Irish Land and the Irish People.*

APPENDIX I.

TABLES OF PRICES IN DUBLIN, 1599-1602.

1599.

Cow	60 shillings	Hen		12 pence
Mutton	10 ,,	Chicken		6 pence
Veal	20 ,,	1 lb. butter		6 pence
Bushel of Wheat	4 ,,	Pig two shillings and 6 pence.		

(Cecil MSS., vol. ix., p. 271).

1601.

Oats   20 shillings a quarter " and not yet good."
Beer   2 pence a wine quart.
Ale   3 pence ,,   ,,
Wheat 50 shillings a quarter.

(C.S.P., Ire., 1600-1, p. 182).

1602.

Wheat	£9 the quarter
Barley malt	43 shillings the barrel
Old malt	22 ,, ,, ,,
Pease	40 shillings the peck
Oats	20 shillings the barrel
Beef	eight pounds the carcase
Mutton	26 shillings the carcase
Veal	29 ,, ,, ,,
Lamb	6 shillings
Pork	30 shillings.

(Leland, History of Ireland, vol. ii., p. 410).

APPENDIX II.

WAGES IN IRELAND IN 1608 AND 1640.

A Note of Rates for Wages of Artificers, Labourers, and Household Servants set down within the County of Tyrone.

(1) All manner of persons being under the age of 50 years, not having to the value of £6 sterling of their own proper goods, shall be compelled to labour for their living. (2) No labourers nor servants shall depart out of one Barony into another, without leave of a Justice of Peace. (3) No persons not having the eighth part of a plough shall keep any servant in their house, but shall labour and do their work themselves. (4) No person shall hire any servant for less term than a year. (5) No servant shall depart from their master without giving a quarter's warning before witness, and at the end of their terms their masters shall give them certificate of their good behaviour, upon pain of 40s. (6) All masters shall pay their servants their wages quarterly. (7) No person shall harbour or relieve any servant being departed from his master without certificate, upon pain of 10s. (8) Every plough holder shall have for wages by the quarter 6s. 8d. sterling, with meat and drink. (9) Every leader of the plough shall have by the quarter 5s., as before. (10) Every beam holder shall have by the quarter 3s. 4d. sterling. (11) A good servant maid by the year 10s. (12) Every young girl serving, rateably. (13) A cowboy for every cow, for the year, 1½d. (14) A cowboy for two heifers, 1d. (15) Every labourer shall be hired by the day, with meat 2d. (16) From Michaelmas to our Ladylady in Lent, with a dinner, 2d. (17) Every labourer without meat, per day, 4d. (18) A master carpenter or mason shall have per day, with meat and drink, 6d. (19) Without meat and drink, 12d. (20) All under carpenters and masons being next to the master per day, with meat and drink, 8d. (22) Every apprentice being able to work well, 2d. (23) For making every plough beam, with meat, 8d. (24) For the best cow-hide, 5s. (25) For the largest pair of broaghs, 9d. (26) For the second sort, 8d. (27) For women's broaghs, 6d. (28) The best plough iron shall be sold for 4s. (29) For making a plough iron, the owner making finding iron, 18d. (30) For the best mending of a plough iron as before, 8d. (31) Every smith shall bring axes, spades, shovels and such necessaries, to the common markets. (32) A weaver shall have for every weavers slatt containing 3 market slatts, 4d. and 8 quarts of meal, of 1,000 or 1,600 a medder of meal, and 1d. (33) For every

IN THE SEVENTEENTH CENTURY.   243

such slatt of 8 or 9 hundred, 4d., and 8 quarts of meal.
(34) For every like slatt of 6 or 7 hundred, 2d. and 4 quarters
of meal. (35) For the best "brakan" weaving after the rate
of the best linen. (36) All other coarse plodding after the
rate of 8 or 9 hundred. (37) For weaving a mantle, a medder
or 2 gallons of meal, and 3d. (38) For weaving the best
caddowe, a medder of meal, and 4d. (39) For weaving of a
jerkin cloth, 2d. (40) For weaving of a trous cloth, 1d.
(41) A cottener for the best mantle, cottened of the best
fashion, his dinner, and 6d. (42) For cottening of a second,
being coarser, his dinner and 4d. (43) For cottening the best
mantle with cards, his dinner, and 4d. (44) For cottening the
best caddow with cards, his dinner, and 6d. (45) For cottening the best caddow with shears, being the best fashion, 8d.
(46) For cottening a jerkin cloth, 2d. (47) For a trous cloth,
1d. (48) Everyone leaving or refusing to work because of
these rates is to be fined, or imprisoned until he be content, 40s.
(49) Every tradesman working at these rates is to have servants to follow his other business.

(C.S.P., Carew, 1603-24, p. 29).

1640—NOVEMBER 30. "A NOTE (IN SIR PHILIP'S HAND) OF THE
HIRED SERVANTS AT CASTLE WARINGE, AND THEIR WAGES
BY THE YEAR."

	£	s.	d.
James Scully, bailiff and overseer	8	0	0
John Farrell, driver, diet and	2	0	0
Alexander Scully, holder	8	0	0
William Read, shepherd, diet and	2	0	0
William Browne, carriage man, diet and	1	5	0
Mary Foulke, diet and	2	0	0
Any (?), diet and	1	10	0
Edmond Blanow, bailiff of the manor	2	0	0
Oliver Birne, diet and	0	6	0
Teige McShane, cowherd	6	0	0
Anthony Geffery, gardener	12	0	0
Nicholas Gorton, weeder	6	13	4
Barnaby Evans, wainman, diet and	2	0	0
Thomas Lawlis, carter, diet and	2	10	0
Murrogh Doole, labourer	7	0	0
Thomas Crowdan, plough carpenter	1	6	0

(Egmont MSS., p. 122-3).

APPENDIX III.

Extract from a Letter from Wentworth to the Lord Treasurer dated 31st January, 1633, proposing a Monopoly of Salt.

" Finally, There remaineth the Business of the Salt, in my Judgment of far the greatest Consequence, and therefore you will pardon me, if I discourse it to your Lordship at large, albeit in the End of a long Dispatch.

The first Ground I shall lay is, the Impossibility of the Patentees furnishing the Quantities of Salt this Kingdom spends yearly by Reason of the Scarcity and Dearness of Fuel, which is a most certain Truth; and for any Man to think to be able in this Country to go through the Work with Peat is a very Mockery, there being not one Summer in ten, which brings Drought and Sun enough to dry and make useful a Piece of that Quantity, which will be spent in making a Proportion requisite. Besides, admit they could, yet the greatest Part of the Consumption is taken up in salting their Fish and Beef, wherein the Salt they shall make here is small and altogether unserviceable, must of Necessity therefore be furnished out of Spain and France.

" The second Ground is upon the Reason of State; for, I am of Opinion, that all Wisdom advises to keep this Kingdom as much subordinate and dependent upon England as is possible, and holding them from the Manufacture of Wool (which, unless otherwise directed, I shall by all Means discourage) and then inforcing them to fetch their Clothing from thence, and to take their Salt from the King (being that which preserves and gives Value to all their native staple Commodities) how can they depart from us without Nakedness and Beggary? Which in itself is so weighty a Consideration as a small Profit should not bear it down.

" The third ground is, the Easiness of making his Majesty sole Merchant: Salt being so perishable a Commodity at Sea, and carrying so great a bulk, as it is not easily to be stolen into the Kingdom; and yet again of so absolute necessity, as it cannot possibly stay upon his Hand, but must be had whether they will or no, and may at all Times be raised in Price so far forth as his Majesty shall judge to stand with Reason and Honour. Witness the Gabelles of Salt in France.

" My fourth and last Ground is Profit: The Corporation of Salters by their new Grant are to give unto the King, upon every Weight containing ten Barrels, 10s. whereof the Medium

IN THE SEVENTEENTH CENTURY. 245

imported yearly, comes to 6,000 Weight, which makes £3,000 the Year; but out of this are to be taken the King's Customs, utterly lost, if the Salt be made within the Kingdom; and those being now set at 2s. the Weight, come to £600, which Loss being deducted, the clear Advantage to the Crown from the Corporation, comes but to £2,400. Besides, that the true Value of Salt being £4 a Weight, the Custom might well be doubled, and consequently the Loss as much more in his Customs. Whereas my Proposition standing with Reason of State, and bringing with it not only a Possibility but indeed a Facility to be effected, shall yet answer a far greater Profit in present, the King's Customs being rather better'd than made worse by it, and be still a fit means in the Hands of a State, to fetch a greater Advantage out of, as Occasion shall advise.

Now the Rate of Salt usually in this Kingdom is between 10s. and 16s. the Barrel: I would set the Medium at 12s. the Barrel, a Rate that would pass within two or three Years without any Man's being sensible of it, and thereout reserve clear to the King 2s. a Barrel for the sole Licence of bringing in and vending this Commodity. And to show that I propound no vain airy Thing, let me have the sole bringing in of Salt, and Liberty to sell at 12s. a Barrel, a Year and a half's time to fit me for the Undertaking, and I will become the King's Farmer so long as I stay in this Kingdom, at £6,000 by Year.

" Nor admits it any Objection, save that you may say, it will leave the Merchant to seek for his Return which he brings now in Salt, and therefore decay the other Trades of the Kingdom: But, besides that the Merchant will be able to find out some other as Beneficial Commodity to make his Return upon, if the Salt be all made in the Kingdom (which certainly in the last Resort is the Aim of the Corporation, or else it is a Business not worth lending an Ear to) it will be far worse that Way too than this Way of mine: For it is still better for Trade, that Salt (which must of Necessity one Way or other Carry Commodity forth to ballance the Value of itself) be still brought in, albeit it pass by another Hand, than when none at all comes from foreign Parts, as in this Case of the Corporation it will fall to be."

(Strafforde's Letters, vol. i., pp. 192-3).

APPENDIX IV.

Ormond's Protest against the Cattle Act.

Letter from the Lord Lieutenant and Council to the King, dated 15th August, 1666.

When I, the Lieutenant, arrived here from England in September, 1665, I was struck by the poverty of the people, which I thought, might lead to public inconveniences. It was, I thought, caused chiefly by the restraint upon the exportation of cattle from Ireland to England, and, by my letters to the Lord Chancellor of England on 18 and 20 September and 11 October and my letters to Lord Arlington of 18 September gave my opinion on the results of that restraint, which was but for some months in the year. Afterwards when I heard of a new bill offered to the House of Commons there for a total restraint, which, it appeared must be total destruction to the Irish people, I consulted the Council. We have now to observe to your Majesty that in July, 1663, we directed the Earl of Anglesey, who was then repairing to Council, to inform your Majesty that people in Ireland were much troubled by a rumour that a restraint would be placed upon their cattle trade. We are assured that he did so; but nevertheless, an Act for encouragement of trade passed in England, by a clause whereof the importation of cattle from Ireland to England was restrained between 1 July and 20 December in each year.

Since then many of the inconveniences and miseries which we early foresaw " are come in like a flood upon this your Majesty's kingdom, whereby the hopeful progress we were in of bringing it to a condition of subsisting of itself and bearing its own charge without burden to your Majesty's revenues in England is frustrated, and notwithstanding all the endeavours we have used to apply remedies to remove or lessen those evils which are befallen your people here they have so totally failed that all the parts of this kindom are (by occasion of the said Act) sank into so sad a degree of poverty and scarcity of money that the misery now fallen upon them is now too sensibly felt by all persons of all qualities and professions throughout the kingdom, insomuch as it has now become an occasion of public grief and discontent in the minds of your people." This trouble coincides with a season of war and may therefore be followed by further inconveniences, as was lately reported to your Majesty through letters to me, the Lieutenant, by the Lord Deputy (Ossory) and Council.

Whilst in these difficulties we were surprised to hear that a bill had passed the Commons in England for a total prohibition of the importation not only of live cattle, but also of beef, pork, bacon and fish, thereby putting us in Ireland in a worse condition than we were in before. We have thought it well to employ some members of our Board to your Majesty on this occasion—namely, the Earl of Cork and Burlington, Lord Treasurer of Ireland, the Earls of Anglesey and Ossory and Viscount Conway, who are instructed to offer reasons which, we think, may move your Majesty to protect your Irish subjects from the ruin which they are threatened. When you have heard their arguments, we have no doubt you will ordain such means of redress for the present grievance and prevention of further mischief as may restore trade and tend to the general relief and satisfaction of your subjects here. To them " we must give this just testimony—that they have upon all occasions since your Majesty's happy restoration manifested a readiness, even beyond their abilities, to advance your Majesty's honour and profit." If, after hearing what is represented to you on behalf of Ireland, your Majesty shall be still prevailed on to pass the Act, we, who could not now keep silent shall acquiesce, in the knowledge that we have done our duty in our office.

We have considered the deficiency of the revenue here. *Details*: An army here cannot even now be paid from our local revenue. This deficiency cannot but be increased by the restraint on the cattle trade and the army must, if it is imposed, be supported by treasure sent out of England. " But when by that restraint on cattle, the discontents of the people shall increase and your revenues be further lessened, then it will be necessary (even for that reason) to increase the army for the better securing of the kingdom. And when all your revenues here are so far short of defraying the charge of the present, how much more short will they fall of paying a greater army. And all those deficits must also be supplied by treasure from time to time to be sent out of England hither or otherwise (which we cannot mention without horror) the kingdom will be in danger to be abandoned and left as a prey for a foreign enemy when it shall be impossible to support your Government here, or to keep in obedience a people universally reduced to poverty and desperation, a consequence of such and so great danger to England as must in such case then require a vast expense of English blood and treasure."

We offer to your Majesty's consideration that, in case the Act pass, we be authorised to call a Parliament in this kingdom and that in order thereunto, we may prepare bills to be transmitted to your Majesty for repealing the statutes enforced in Ireland prohibiting the export of sheep, wool and other commodities into any part of the world but England. Such bills, if passed, would in some degree mitigate the extremities falling on this country by the stoppage of the cattle trade to England. "And in the meantime that, after the passing of the act and until your Majesty shall think fit to call a Parliament, you may be pleased to give us leave that we may by Act of Council and proclamations, inhibit the importation into this kingdom of such commodities as may take away from us even that little base and foreign coin which now supports the small trade and commerce we have."

We confess that we deeply apprehend your Majesty may be in some difficulty as to what course to take in this matter when the whole of one of your kingdom's demand one course and some people in the other demand another. The consideration of this difficulty "doth much perplex your servants." But in case the benefit conferred by such prohibition on some of your people in England be but little, and the loss in which it will involve Ireland universal and great, we think that the greater evil which will fall on Ireland, and that too attended with dangerous consequences even to England, should outweigh the small advantages which certain Englishmen gain by the prohibition.

We submit all these matters with humility to your Majesty's judgment, "beseeching Almighty God so to guide your counsel and resolutions, as in all things, so particularly in this, as may be safe to your Royal person, healthful to all your kingdoms, satisfactory to all your people and profitable towards preventing the underhand working of your enemies."

(C.S.P., Ire., 1666-9, pp. 183-5).

APPENDIX V.

CONTEMPORARY ACCOUNTS OF DAMAGE INFLICTED ON IRELAND BY THE CATTLE ACTS.

The Lord Lieutenant and Council to the King, 9th Feb., 1666-7.

Our duty compelled us to represent to your Majesty the prejudice which Ireland sustained when the export of Irish cattle was prohibited annually for a part of each year. Later when a bill for the total prohibition was under consideration, we, on August 15 last, represented what a fatal effect the bill, if passed, would produce in Ireland.

Hearing that this bill is past, we must at once inform your Majesty that " if by your royal goodness and care some speedy and effectual remedies be not extended to your subjects of this kingdom it will too probably be above the best endeavours of us, your majesty's servants to preserve it in that condition in which we have laboured and still labour to do." The war and the late Act passed in England " for encouragement of trade," prevent our treating with foreign countries and the plantations and this new law forbids our chief trade with England.

Your Majesty's revenues do fall in proportion to the decay of your subjects' traffic. Some who cannot now live by their labour maintain themselves by the spoils of others, and we have too much cause to believe the numbers of such bad people will daily increase as their wants do, whereby while there is need (both at home and from abroad) to augment your Majesty's army, the treasure which should pay it lessens.

Even while we had peace, the revenue here was not sufficient to pay the civil and military establishments; so that " it is evidence that from henceforth this kingdom and people must suffer a destructive consumption unless from your Majesty's goodness they receive a sudden relief."

At present, however, we desire to suggest remedies. We have carefully considered these; and they are of two sorts those for which new laws are necessary here and those which may be obtained by your Majesty's own grant. We reserve the legislative remedies until your Majesty think fit to call a Parliament here, and at present only mention those which we consider that your Majesty has power to grant. We speak earnestly on this matter only from necessity; because, though we, by the blessing of God, not yet know miseries of invasion or rebellion, we struggle under difficulties which, if not speedily removed, will as certainly, though not so speedily, ruin us.

Upon this occasion we thought well to see what the " most expert merchants in this kingdom would offer on these three essential points, first, the support of trade and manufacture, secondly the bringing of money into this kingdom, and thirdly hindering the carrying out of that little left in it. And although it was readily offered that nothing could more certainly and expeditiously answer the first two ends than the obtaining your Majesty's leave to transport the wool of this kingdom into foreign ports, nor reach the third end which we propose than to prohibit the importation of English commodities into Ireland which carry away the ready money of this kindom yet we, your Majesty's servants, apprehending not only that the detriment we suffered by the late Act against the Irish cattle may be found ere long in the practice to be as prejudicial to the body of the people of England as to your Majesty and your subjects of this kingdom, but also that humble desires of that nature might possibly, if granted, be judged as much a prejudice to England as a benefit to Ireland, and being likewise earnestly desirous to evidence that no hardships to be appointed can anyway diminish the due regard we have and ought to have of the good of your Majesty's kingdom and people of England (the prosperity of which we wish as heartily as our own) we have wholly employed our thoughts to find out such remedies in our sad condition as may, in the consequence of them, bring no detriment to England or any of your Majesty's other dominions and may be disadvantageous to those who have now the misfortune to be your Majesty's enemies."

We therefore ask your Majesty to licence the transport of the commodities of the kingdom (wool excepted) into France, Holland, Norway and Denmark and any other places, though at enmity with your Majesty, and for letters of your Majesty's fleet and to all privateers having commission from your Majesty, etc., ordering that all ships coming here from there or going there from here, and having passport from the Lord Lieutenant (provided they do not carry any more arms than are necessary to defend themselves) may pass and repass between Ireland and those countries with their merchandize, whereby the trade of Ireland may be kept alive and the revenue of Ireland kept from sinking. We must either have leave to trade with your enemies or else we must be supported by despatch of money from England here, as was usual in old days. If this leave is given we will "tie" all merchants who are licensed to trade with the enemy to bring back a reasonable proportion in cash or bullion of the product of the native commodity they export which will be more considerably prejudicial to your enemies than our native commodities will be advantageous to them; for we cannot subsist without money and they may live without our merchandize. Nor will it be a little advantageous to England that, by this grace from your Majesty to Ireland, will in the consequence of it get most of the money or bullion which our commodities will bring in. For, no prohibition being yet laid on the importation of English commodities into Ireland, and those commodities being to be paid for in money (for the act against cattle renders money the only thing with which we can buy what England sends us) England will in effect have the advantage of the grace your Majesty shall herein extend to Ireland.

For the encouragement of our trade we also ask that your Majesty will command the Governors of all your American islands and plantations, and all others concerned, not to put in execution against the subjects of Ireland, the Act of the 15th year of your reign "for the encouragement of trade." But that Irishmen may have liberty to treat with all those Islands, etc., as they had before that Act was passed. This trade does not bring us in ready money but it does take off the manufacture of Ireland "to which the people must addict themselves, now that the cattle trade is stopped. Manufacture is a work so much beside the genius if not the inclination of the people of Ireland and by encouragement is at first desirable and necessary."

We also ask that a considerable part of your fleet may be victualled in Ireland. This we hope will be as much to the advantage of the fleet as to them of Ireland. Probably the prohibition of Irish cattle will cause the price of beef to rise in England as it falls in Ireland; and we know not but whether the contractors for victualling the navy may not seek a new contract owing to this change. And although it may be said that not much Irish beef or pork was consumed in the navy by reason of its being short in goodness of the English, yet certainly the amount of cattle exported into England did moderate the price of beef and pork, which were consumed in the navy. Moreover, we do not think that the victuallers need fear the use of such a supply, for although now Irish beef and pork are reckoned as inferior owing to the want of skill of those who kill and sale them, we believe that if the contractors employ "knowing factors of their own," here to prepare the supply, beef and pork might certainly be supplied of as good a quality as that in use in the navy now. The constant demand will moreover encourage "many of the natives" to a better breeding of cattle, and not to kill them for sale till they are six or seven years old. "When none but good cattle are a commodity and such have a good vent as well in foreign ports as for the fleet, their profit will invite them to what nothing else will or at least hereto could."

We must, however, at once have £60,000 and that in coin, otherwise all these remedies, even if granted, will be but as food brought to those who are first starved. If not in English coin it should come in such foreign coin as is current amongst us; for since the late Act will change the whole way of livelihood of the people of this kingdom, and necessitate them from the breeding of cattle to fall to manufacture (in which we shall most encourage the linen, as being no way opposed to the manufacture of England) it will be highly useful and necessary that ready money might be sent us, which being paid the army will be dispersed amongst the people and so fit them for and encourage them in this way of living.

The granting of these requests will be a large evidence of your Majesty's royal bounty and goodness to your Irish subjects and they sorely need such comfort, " for it is but too commonly discoursed amongst them upon the reading of those acts lately past there that it is necessary the English of Ireland should return to their native country, since England in a manner seems desirous as it were to cut off their dependence on it."

If England abound with their own breed of cattle, the Scotch cattle as well as the Irish would have been forbidden. If it does not, why are not our cattle as well admitted as theirs, when by carrying them from England they not only yield considerable duties to the kingdom here and there, but also the product of those cattle in England yield the like duties at their return to us to his Majesty in both kingdoms, which the Scotch cattle in neither case do. In effect the Act will carry the Scotch cattle into England and the Irish cattle into Scotland.

If England had not an aim to draw the English of Ireland into England they would not have made that law which does forbid them to treat with their native country, and that in a time when the war forbids them trade in all others, but have given them at least some competent forewarning to have set themselves to manufacture before the trade of cattle was prohibited, and not necessitated them to such total change of the way of their livelihood in a juncture when war disables them to get artificers to teach them another or to vend their manufacturers when they should have wrought them and made then fit for sale..

It is in these and such other like discourses that the afflicted people of Ireland give some vent to their grief, and therefore, in so general a calamity and consternation, we must humbly beg your Majesty to commiserate their condition, " and by your goodness keep them from despair. Nothing can so effectively do this as your granting them these few particulars, which we humbly beg of your Majesty for them.

" And although we cannot say that these graces, when granted, will prove a remedy proportionable to the disease, yet we will improve them to the utmost of our power and support and maintain your Majesty's Government in this kingdom and to relieve and comfort your Majesty's subjects in it."

(C.S.P., Ire., 1666-9, pp. 289-93).

The Lord Lieutenant to Secretary Arlington, 16th Feb., 1666-7.

The ruin which the late Act against Irish cattle brought upon some particular persons who had made provision of cattle to transport in confidence the Act would not pass at all, or not so soon as it did, has raised their ingenuity or despair to such a degree that some ship loads of cattle are transported into England since the Act came printed hither. By what trick or composition it is they hope to evade the penalty imposed I do not well know, and how either will succeed with them. But I presume those who drove on that Act with so much earnestness and have therefore undergone the hard opinion of many, and a gentle reprimand from the King, will be highly displeased to find so early an attempt to elude what they had so hardly obtained; and what inconvenience that may produce if the Parliament is still sitting, or shortly to sit, I cannot so well answer, but that I would have stopped this exportation of our cattle till the king's pleasure be known if I could have found any law or conscience for it, and I am not willing to be unreasonable for the inconvenience such prohibition might have produced here where I am more immediately applied to the King's service. Yet I hold it necessary to give your lordship this intimation though you may and are like to receive it more warmly from others. I confess I am sorry this transportation of cattle will not suffer those who believe the falling of rents there proceeds from it to be undeceived, as I am confident in a short time they would be, and will be if the prohibition take place. In the meantime if the trick will serve I know not but it may do as well for the whole year as for these months.

(C.P.S., Ire., 1666-9, p. 302).

COPY OF MEMORIAL PRESENTED TO THE KING ON FEBRUARY 17TH, 1666-7, BY THE EARLS OF ANGLESEY AND BURLINGTON AND VISCOUNT CONWAY ON BEHALF OF IRELAND.

Since the Act against Irish cattle is passed, we beg leave to discharge a trust laid upon us by the Lord Lieutenant and Council of Ireland by presenting the present condition of Ireland and the great evils which must follow. We shall also try to offer remedies, which your Majesty can apply.

The total prohibition of importing Irish cattle " both here and in Scotland," following the partial prohibition which has lasted for some years, has reduced people here to desperate poverty. " Many of those who cannot find a livelihood by the the trade of cattle, wherein generally the Irish employed themselves, are already gone into actual rebellion, burning and spoiling the English, and it is no ways to be doubted the necessities and poverty of the generality will daily increase their number, which will desist and disappoint all payments to your Majesty, weaken the hands of your good subjects, and may invite and facilitate foreign invasion."

For remedy, it is humbly proposed that all restraints upon the exportation of commodities of the growth and manufacture of Ireland to any foreign parts may be taken off, and in his time of war by permission of His Royal Highness, the Lord Lieutenant or Deputy's pass may secure any ships loaden in Ireland to France, Holland, Denmark, "any part of the world in amity with your Majesty where free trade shall be allowed them or connived at, and that the said passes do secured ships whilst going and returning."

It is also proposed that "to preserve the little money which is left in Ireland for the carrying on the trade there and paying your Majesty's Army, and to repress the luxury and humour of the people after outlandish commodities, your Chief Governor and Council may have direction to publish a proclamation or Act of State prohibiting the importation of all commodities of the growth or manufacture of Scotland till such times as the restraint upon Irish cattle and commodities in Scotland be taken off." Ireland should have the same liberty in the matter as Scotland has taken. It is also asked that the Lord Lieutenant and Council may have direction to consider the best course "to regulate the disordinate import of such outlandish commodities which the kingdom may well live without."

"Lastly, since it is apparent that your Majesty's revenue will extremely dimish henceforth in Ireland, and their dangers and their difficulties increase that would preserve that kingdom to your Majesty," we humbly offer it as absolutely necessary that £50,000 be at once sent in specie to Ireland to pay the army and Civil List and to answer any emergency occasioned by "the disjointed condition of that kingdom."

(C.S.P., Ire., 1666-9, p. 303).

APPENDIX VI.

PROPOSALS MADE BY SIR WILLIAM TEMPLE IN HIS ESSAY ON THE TRADE OF IRELAND FOR THE REGULATION OF THE PROVISION TRADE.

No great or useful thing is to be achieved without difficulties; and therefore what may be ruled against this proposal ought not to discourage the attempt of it. First, the Statutes against that barbarous custom of ploughing by the tail ought to be renewed, and upon absolute forfeitures instead of penalties; the constant and easy compositions whereof, have proved rather a letting than a forbidding it. Now if this were wholly disused, the harness for horses being dearer than for oxen, the Irish would turn their draught to the last wherever they have hitherto used the ploughing by the tail. Next, a standard might be made under which no horse should be used for draught; this would not only enlarge the breed of horses, but make way for the use of oxen, because they would be cheaper got than large good horses which could not be wintered like garrans, without housing or fodder. And lastly, a tax might be laid upon every horse of draught throughout the kingdom; which besides the main use here included would increase the King's revenue, by one of the easiest ways that is anywhere in use.

For the miscarriages mentioned in the making up of those several commodities for foreign markets, they must likewise be remitted by severe loss, or else the matters of the commodities themselves will not serve to bring them in credit, upon which all trade turns. First the ports out of which such commodities shall be shipped may be restrained to a certain number, such as lie most convenient for the vent of the English Provinces and such as either are already, or are capable of being made regular Corporations. Whatever of them shall be carried out of any other Port shall be penal both to the merchants that delivers, and to the master that receives them. In the Ports allowed shall be published Rules agreed on by the skilfullest merchants in those ways to be observed in the making up of all such as are intended for foreign transportation and declaring that what is not found agreeable to those rules shall not be suffered to go out. Two officers may be appointed to be chosen for three years by the body of the corporation, whose business shall be to inspect all barrels of beef, tallow and butter and packs of hides and put to them the seal or mark of the corporation, without which none shall be suffered to go abroad, nor shall this mark be affixed to any parcels by those officers but such as they have viewed, and found agreeable to the rules set forth for that purpose. Whereof one ought to be certain that every barrel be of the same constant weight, or something over. If this were observed for a small course of time, under certain marks the credit of them both as to quality and weight would rise to that degree, that the barrels or packs would go off in the markets they use abroad, upon sight of the mark, like silver plate upon sight of the City's mark where it is made.

The great difficulty will lie in the good execution of the offices; but the interest of such Corporations lying so deep in the credit of their mark, will make emulations among them, everyone vying to raise their own as high as they can, and this will make them careful in the choice of men, fit for that turn. Besides the office ought to be made beneficial to a good degree, by a certain fee upon every seal, and yet the office ought to be forfeited upon every miscarriage of the officer, which shall be judged so by the chief magistrates of the town, and thereupon a new election be made by the Body of the Corporation.

APPENDIX VII.

Sir William Temple's Opinion of the Cattle Acts from the Essay on the Trade of Ireland, 1673.

Cattle for exportation are bullocks and horses; and of one or other of these kinds the country seems to be full stocked, no ground that I hear of being untenanted. The two first seem sufficiently improved now, kinds as well as the number, most of both being of the English breeding. And though it were better for the country if the breed of horses being lessened, may work for that of increasing sheep, and great cattle; yet it seems indifferent which of these two were most turned to, and that will be regulated by the liberty or restraint of carrying like cattle into England. When the passage is open, land will be turned most to great cattle; when shut to sheep as it is at present; though I am not of opinion it can last, because that Act seems to have been carried rather by the interests of particular Counties in England than by that of the whole, which in my opinion must be evidently a loser by it. For first, the freight of all cattle that were brought over being in English vessels, was so much clear gain to England, and this was one with another a third, or at least a fourth part of the price. Then from going over young and very cheap to the first market, made them double the price by one year's feeding, which was the greatest improvement to be made on our dry pasture land in England. The trade of hides and tallow or else of leather, was mightily advanced in England, which will be beaten down in Foreign markets by Ireland, if they come to kill all their cattle at home. The young Irish cattle served for the common consumption in England, while their own large old fat cattle went into the barrel for the foreign trade in which Irish beef had in a manner no part, though by the continuance of this restraint it will be forced upon improvement, and come to share with England in the beef trade abroad. Grounds were turned much in England from breeding, either to feeding or dairy, and this advanced the trade of English butter, which will be extremely beaten down when Ireland turns to it too (and in the way of English housewifery, as it has done a great deal since the restraint upon cattle), and lastly, whereas Ireland had before very little trade but with England, and with the money for their cattle bought all the commodities there which they wanted; by this restraint they are forced to seek a foreign market; and where they fell, they will be sure to buy too; and of the foreign merchandize which they had before from Bristol, Chester, and London, they will have in time from Rouen, Amsterdam, Lisbon and the Streights. As for the causes of the decay of rents in England, which made the occasion of that Act, they were to be found in the want of people, in the mighty consumption of foreign commodities among the better sort, and in the higher way of living among all, and not in this prosperity of Irish cattle, which would have been complained of in former times, if it had been found a prejudice to England. Besides, the rents have been far from increasing since; and though that may be by other accidents, yet as to what concerns Ireland, it comes all to one, unless wool be forbidden as well as cattle; for the less cattle comes from thence, there comes the more wool, which goes as far as t'other towards beating down the price of pasture lands in England; and yet the transportation of wool cannot be forbidden, since that would force the Irish wool, either by stealth into foreign markets, else in cloth to the advance of that manufacture; either of which would bring a sudden decay upon the principal branch of the English trade.

APPENDIX VIII.

Memorandum Urging the English Parliament to Repeal the Cattle Acts.

Reasons offered to the consideration of Parliament for remitting the prohibition on the importation of Irish cattle :—

1. It has proved very prejudicial to the Customs revenue, not only the duty of cattle, etc., imported from Ireland being lost, but also of the Customs formerly paid on goods imported into England and sent to Ireland in return for their cattle

which formerly paid a custom on importation, another on exportation, and a third in Ireland.

2. It has greatly prejudiced the landowners in England. (i.) Because breeding lands in England are not able to raise a stock for the feeding. (ii.) It makes breeders impose a greater rate for their lean beasts than they could be sold for when fatted, which makes feeding lands worth little or nothing. (iii.) It has transferred most of the victualling both for home consumption, foreign trade and naval provisions from England to Ireland, and the places where Ireland sends them, so, though land cattle be dearer than before the prohibitions, fat cattle are cheaper, because we have lost the former consumption of them.

3. It is destructive to navigation and trade. (i.) To Navigation.—When they were imported, at least 300 or 400 ships were constantly employed in that trade, but since the prohibition the whole of the trade is managed in foreign bottoms. (ii.) To Trade—because foreigners who formerly victualled here now victual in Ireland where they have beef at 12s. a barrel, which is $2\frac{1}{2}$ cwt.; in England 23s. and 24s. a cwt. is paid. (iii.) It will not be difficult to prove that Holland has during this war been supplied with their naval provisions out of the King's own Dominions for a fourth of the price he has to pay for his; let everyone then judge with what disadvantage he has gone to war with his enemies; having provisions so much cheaper, they can sail for less freight and wages, and so have great advantage of us as to trade and may undersell us. To prevent this, the English now, when they sail on long voyages, only take a month's or six weeks' provisions here, because in Ireland, Spain or elsewhere they touch at, they get supplied with Irish provisions much cheaper than in England. (iii.) The Irish do not take money for the cattle, but English manufactures, by which the poor were employed and spent their earnings on English goods and manufactures, thus keeping up their price, but now the Irish fetch the goods they want from beyond the seas, so that the traders in Lancashire and Cheshire and elsewhere breeding lands lie, lose more for want of consumption of the manufactures there, than the breeders get by the price of land cattle.

4. The prohibition has made Ireland lessen their great cattle and increase their sheep, so that they have prodigious quantities of wool, which with their hives and tallow is mischievous to England in three ways. (i.) By sending quantities of wool beyond seas unmanufactured, by manufacturing which foreigners grow rich, while the poor here starve for want of

the work they formerly had for foreign consumption. (ii.) By vast quantities of wool sent to England, bringing down the price of our own. Similar mischiefs attend the importation of their hides and tallow. (iii.) By setting up woollen manufactories in Ireland, where having wool, hides and tallow cheaper than we, and all sorts of provisions at a much less rate, they must have workmen for half the price. If then the raw material and the manufacture be cheaper there than here, what shall hinder not only their making woollen and leather manufactures for their own use, but also supplying foreign markets and then the staple trade and commodities of England will necessarily be undermined, and the many hundred thousands who depend on the manufactories thereof will be reduced to beggary, and England want the consumption of the provisions and manufactures they spent when employed, which must bring down their prices and consequently that of land?

This prohibition has made Ireland incapable of trade with England because they cannot pay for what they buy unless they send over money in specie, which tends to ruin that Kingdom, or return it by bills of exchange, which cost 15 or 16 per cent.; and that is double the advantage the trader gets here by the sale of their commodity, which also forces those who live here, whose estates lie in Ireland, to retrench a sixth of their expenses of the land, which is a further hindrance to the consumption of goods grown or manufactured here.

It has undone many eminent traders, haberdashers, etc., in London. For, besides the staple commodities sent thither, when fashions were out here, they went to Ireland in return for their cattle and were as good as new, for want of which utterance, by reason of the often change of fashions here, many tradesmen have been undone.

It is likely to prove fatal to the English fishery, for Ireland being thereby put upon industry and parts of it lying nearer to France, Spain and Italy, than England does, they having salt from France and cask in Ireland cheaper than we have in England, and provisions and wages being cheaper there, having set up a fishery trade there whence they need but one wind to carry them to the Foreign market and catching their fish six weeks before we take in England and lying so many leagues nearer the market, what hinders them from being at market sooner than and cheaper than we can?

By reason of the loss of our manufactures our people are removed and removing thither, as combers, weavers, etc., which will prove a great advantage to them and a greater disadvantage to England than bringing over their cattle.

If the surcharge on Irish cattle of 20s. a head for customs, freight, etc., be not thought sufficient, it is left to the wisdom of Parliament to settle it, so as to be least prejudicial to England.

The riches of a nation arising from the labourer, artificer and manufacturer, from their labour money is first raised to pay the tenant and through him the landlord; now to take the labourer is to stifle the riches of the nation in embryo, and how much the prohibition has in reality or pretence done this is to be considered. Manufactures and manufacturers are much lessened thereby, the exportation from England to Ireland before the prohibition being £204,000 per annum, which will now scarce be found above a tenth thereof, which results in a double prejudice.

1. The manufactures of the land and things of the growth of this nation are much lessened.

2. The consumption of the Nation is lessened, and the labourers left cannot feed so well, since provisions are dearer and the most minute addition on that amounts to a vast bulk in the total. For, supposing the people of England to be 8,000,000 and to stand one with another 20s. worth of flesh meat at an average price of 2d. per lb., adding on $\frac{1}{4}$d. per pound to the price amounts to a million of money spent more than before, whereas the cattle from Ireland did not amount to above £80,000 or £90,000 per annum or thereabouts; and since those who labour are in proportion the body of the Kingdom, and generally beef and mutton eaters the dearer that is the more they spend on meat and the less they have to pay to the tenants and they to the landlords. Further, if riches consist chiefly in the work of hands and no kingdom is rich for what it consumes in itself, but in what it furnishes abroad, then breeding cattle in England, which requires but few things, tends to disable the land, and so to the abatement of the riches thereof, and to encourage that with a preference to our manufacturers will be quite contrary to the policy of former times, which provides laws principally enjoining tillage and manufactory, as the best means to increase the yeomanry, who were deemed the strength and riches of the kingdom."

(C.S.P., Dom., 1673-5, pp. 166-9).

APPENDIX IX.

AN ACT FOR THE ADVANCEMENT AND IMPROVEMENT OF TRADE, AND FOR ENCOURAGEMENT AND INCREASE OF SHIPPING, AND NAVIGATION.

" Whereas, this Kingdom of Ireland, for its good Situation, commodious Harbours, and great Quantity of Goods, the Growth, Product, and Manufactury thereof is, and standeth very fit and convenient for Trade and Commerce with most Nations, Kingdoms and Plantations; and several Laws, Statutes and Ordinances, having heretofore been made and enacted, and time to time, prohibiting and disabling the King's Subjects of this Realm, to export, or carry out of this Kingdom, unto any other the King's Island, Plantations, or Colonies, in Asia, Africa, or America, several of the Goods, Wares, Merchandizes, and Commodities of this Nation; or to import into this Kingdom, the Goods or Merchandizes of the said Plantations, Colonies and Islands, without landing and discharging in England, Wales, or the Town of Berwick upon Tweed, under great Penalties and Forfeitures not only to the Decay of the King's Revenue, but also to the very great prejudice and disadvantage of all the Inhabitants in this Kingdom, as well Subjects as Strangers; and which hath in a high measure contributed to impoverish this Kingdom, and discouraged several Merchants, Traders, and Artificers, to come from abroad, and dwell, and trade here; And Whereas, the Increase of Shipping, and the Encouragement of Navigation, under the good Providence of God, and the careful Protection of his sacred Majesty, are the best and fittest Means and Foundations, whereon the Wealth, Safety and Strength of this Island and Kingdom, may be built and established. Be it therefore Enacted, by the King's most Excellent Majesty, with the Advice and Consent of the Lords Spiritual and Temporal, and Commons in this present Parliament Assembled; and by the Authority of the same, that it shall, and may be lawful, to and for his Majesty's Subjects of this Realm of Ireland, and to and for every other Person and Persons, of what Nation soever residing and inhabiting here, during the time of such Residence, freely to trade into, and from all and every his Majesty's Plantations, Colonies and Islands, in Asia, Africa and America, and to export from this Kingdom, and carry unto all and every the said Plantations, Colonies, and Islands, and there sell, dispose of, and barter all sorts of Goods, Wares, Merchandizes and Commodities, as well of the Growth, Product, or Manufactury of this Kingdom, as of any other part of Europe, commonly called European Goods, and import, and bring into this Kingdom of Ireland, all sorts of Goods, Wares, Merchandizes, and Commodities of the Growth, Produce, or Manufactury of all or any the said Islands, Colonies and Plantations, without being obliged to land or unload in England, Wales, or the Town of Berwick upon Tweed, or entering all or any such Goods, Wares, or Merchandizes there, but as hereinafter is expressed, and without being obliged upon Shipping, or taking on Board, in the said Plantations, Colonies, or Islands, the said Commodities, to enter into any Bond, to bring the said Goods into England, Wales, or Town of Berwick upon Tweed, and to unload and put the same on shore, Any Act, Statute, Ordinance, Law, Sentence, or Judgment, at any time heretofore made, given, or in force, to the contrary notwithstanding: Provided always, That the Master or Owner of all and every such Ship and Ships, Vessel or Vessels, so trading from this Kingdom, unto all or any the said Islands, Colonies or Plantations, his or their Agents or Factors shall, and do before such Ship or Ships, Vessel or Vessels, sail from any part of this Kingdom towards the said Islands, Colonies or Plantations, perfect and enter into bond, with one sufficient Security, to the use of the King, and to be perfected to the Collector, or chief Custom-house Officer, of such Port or Place, whence such Ship or Vessel is to sail, in such a reasonable Sum, as such Collector, or Custom-house Officer shall require, Regard being had to the Value of such Cargoe, as the said Ship or Vessel shall export, with condition to bring the Goods, Wares, and Merchandizes, which such Ship or Vessel shall take in, at all or any the said Plantations, Colonies, or Islands, into England, Ireland, Wales, or the Town of Berwick upon Tweed, and to no other Place, and there to abode and put the same on shoar, the danger of the Seas only Excepted.

"(2) Be it likewise enacted, by the Authority aforesaid, That all Goods and Merchandizes whatsoever, which shall be carried, conveyed, or exported out of this Kingdom of Ireland, to the said Islands, Colonies, and Plantations, shall be liable and pay to the King's Majesty his Heirs and Successors, in the said Islands, Plantations and Colonies, the same or so much Customs, Excise, or other Duties, as the like Goods or Merchandizes being exported out of England, into all, or any the said Plantations, Colonies, or Islands, and all Goods or Merchandizes imported into the Kingdom out of all or any the said Islands, Colonies, and Plantations (Tobacco and Sugar only excepted) shall pay in this Kingdom to the use of the King's Majesty, his Heirs, and Successors, the same or like Duties, Custom and Excise, and no more or other, and in such manner, and at such time, and subject to such Penalties and Forfeitures, for Non-Entry, Undue Entry, or Non-Payment of Duties, as in the like Acts of Parliament made in this Kingdom, in the Fourteenth and Fifteenth Years of Reign of the late King Charles the Second; the One, Entituled, An Act for Settling the Subsidy or Poundage, and Granting a Subsidy of Tunnage, and other Sums of Money unto His Royal Majesty, his Heirs and Successors; the same to be paid upon Merchandize, imported and exported into or out of the Kingdom of Ireland, according to a Book of Rates hereunto annexed; and the other, Entituled, An Act for the Settling of the Excise, or new Im-Post, upon His Majesty, His Heirs and Successors, according to the Book of Rates, therein inserted, and as in the said Book of Rates, and as in the Rules, Orders, and Directions, to the said Acts and Books of Rates annexed, are contained and specified.

" (3) And Whereas, the Duties, and Custom, and Excise on Tobacco, of the King's Majesty's Plantations, imported into this Kingdom, amount to no more according to the said two late Acts of Parliament in this Kingdom, and Books of Rates to them annexed, but to Two Pence per Pound, which is too small a Duty; Be it therefore Enacted, by the Authority aforesaid, that all Tobacco of the Growth or Product of all or any His Majesty's New Plantations or Islands, or any Plantations belonging to His most Christian Majesty, imported into this Kingdom, out of all or any the said Plantations and Islands, shall from and after the Eighteenth Day of July, 1689, be charged, and lyable to pay unto his majesty, his Heirs, and Successors, the sum of Five Pence Stel. for each Pound, Custom, and Excise, (that is to say Twopence for each Pound, Custom, and threepence for each Pound Excise, and no more); Provided always, That Spanish and Brazil Tobacco, shall pay the same Duty of Custom and Excise, as formerly; and that likewise, Tobacco of that Growth or Product of the King's Plantations, or any of the Foreign Plantations belonging to his Most Christian Majesty, imported into this Kingdom out of England, or any other part of Europe at any time, from and after the Eighteenth Day of July, 1689, shall pay and satisfy unto the King's Majesty, his Heirs and Successors, the Sum of two Pence, Sterl. Custom, for, and out of each and every Pound, and the sum of Two Pence half-penny, Sterl. Excise, for, and out of each Pound and no more. And, That **Sugars,**

Indicoe, Logwood, imported into this Kingdom out of England, shall pay, and satisfy unto the King's Majesty, his Heirs and Successors (viz.), white Sugar coming from England, Ten shillings Custom, and Ten shillings Excise for every hundred weight, and no more; brown Sugar, the Sum of Two shillings and Six Pence, Sterl. Custom, and the like Sum of Two shillings and Six Pence Sterl. Excise for each hundred weight, and no more; Indicoe, the sum of Two Pence per Pound Excise, and Two Pence Custom for each Pound and no more; and Log-wood, Five Shillings Sterl. Excise and Five Shillings Sterl. Custom, for each hundred weight, and no more; The said Duties, Customs and Excise to be paid in such manner, and under such Pains and Forfeitures, and with such Allowances as in the Aforesaid Two Acts and Books of Rates, Orders, and Directions are expressed and contained.

" (4) And for the further Encouragement and Advance of the said Plantation Trade, and for Maintaining a greater and more firm Correspondence and Kindness between the Subjects of this Kingdom and Planters, and Inhabitants of the said Plantations and Islands: Be it enacted, by the Authority aforesaid, That whatsoever Goods or Commodities of the Growth, Product, or Manufactury of the said Islands or Plantations, shall be at any time hereafter Unloaded, or Landed in any Part of this Kingdom, and shall pay or secure to be paid, the Customs, Duties, and Excise on the said Goods, due and payable, that at any time hereafter, within the space of One whole Year, to commence from the day of such Landing, it shall and may be lawful to and for the Merchant, Owner or Proprietor of such Goods and Commodities, his or their Agents or Factors, to export and carry out of this Kingdom into any other Nation, Dominion, or Country, such and so much of the said Goods and Commodities so landed, as he or they shall think fit; and that upon such Exportation the whole Excise of such Goods, which was before paid, or secured to be paid for the same, and one half of the Custom of the said Goods before paid or secured to be paid, shall be re-paid or allowed to such Merchant Owner, Proprietor, his or their Factors or Agents so exporting, and that within twenty Days next and immediately ensuing the Date and Time of such Exportation, Tobacco only excepted."

APPENDIX X.

PROPOSAL OF THOMAS SHERIDAN, ON BEHALF OF HIMSELF AND OTHERS, FOR BUYING ALL THE WOOL OF IRELAND FOR 21 YEARS AND FOR ITS TRANSPORTATION INTO ENGLAND ONLY, TO COMMENCE FROM 1ST MAY, 1674.

The exportation of the wool of Ireland is prohibited. This was done both to hinder foreigners from carrying it away, and to necessitate the Irish to manufacture it at home (which cannot be done without prejudicing England by making any other place the staple for woollen manufactures), yet the Chief Governors have, almost ever since these laws were enacted, given licenses to transport the wool into England only, under certain conditions. However, many merchants have by stealth transported wool into France and Holland, whereby the English woollen manufactories have been much prejudiced and the foreign advanced, for the exporters, if they succeed one voyage in three, were considerable gainers, so much were the rates for wool higher in those foreign parts than in England. Forbidding Irish cattle being brought into England necessitated people there to leave off much of their breed of black cattle, and fall into the breeding of sheep, whereby that kingdom has much more wool than it ever had before, which is likely to increase every year, so, if some speedy and effectual regulation be not made in the Irish wool trade (the manufacturers there not being able to employ one-fourth of it), persons will rather transport it to foreign countries, and run the risk of the law, if taken, than make it a drug at home. Of late the best sorts of wool, which used before the war to yield 10s. a stone in Ireland, when brought to the port to be shipped, have not yielded 7s. to the owners, and proportionably for the less fine. For all the wool masters are either tenants or such proprietors as stock their own lands with sheep, and not being able (as to most of them) to send their wool to England to the best markets, are obliged to sell it to the Irish merchants, who confederate to buy it at their own rates, which are so low that the tenants are not able to live, to sell often to such buyers, nor able to undergo the charge (which is at least 3s. a stone, besides the hazard of the sea) and attendance of carrying their wool to England. The proprietors also who have flocks of sheep are generally subjected to the like inconveniences, and therefore must come to the merchants or keep their wool in their own hands, and are often forced to barter it at their unconscionable rates, whereby also the Irish wool buyers, having it so cheap, may gainfully undersell the English woolsellers, whereby both the freeholders and tenants of England and Ireland, who are shipmasters visibly decay, and the combining merchants only increase, whereas if it bore a good rate in Ireland, the shipmasters of Ireland as well as England would be the better, and the inducements to transport Irish wool to foreign parts would be in great measure removed.

It were therefore to be wished, since none can transport wool out of Ireland to England without a licence, that his Majesty would commission some honest and knowing persons to buy up all the wool that comes to the ports at honest and equal rates between buyer and seller, viz., for the combing wool of Tipperary, Cork, Limerick and Clare 10/6 a stone, the clothing wool being of two sorts, for the best of those counties 7/6; for the worse 6/6; for the combing wool of Leinster and Waterford 10/-; for the best clothing 7/-; for the worst 6/-; for combing wool of Ulster County and Kerry 9/6; best clothing 6/6; worst 5/6; for fell wool of Munster, long 9/6; short 6/6 for long fell of Leinster and Waterford 9/-, and for short 6/-; for the long fell to Ulster county and Kerry 8/6; short 5/6; or, if the people do rather, they will give for the best wool 10/-; for the second 8/6 and the short 7/-, according to its quality, and for fell wool, long 9/-, short 6/-; according to depth or fineness of the staple, and shall pay ready money on the wool being weighed and delivered in Dublin, Cork, or Waterford, which three ports lie most conveniently for all the wool of Ireland, and shall further allow 2d. a stone for every stone brought above 50 miles to any of the said ports, which rates besides the payment of ready money, are much higher than those for several years past, though they are often at 3, 6, 9, and 12 months' time for payment, and are greater than will ever be given by the trade merchants, and those rates are not to be imposed on the seller, though such as both seller and buyer may thrive by.

By the nearest computation the stock necessary to carry on this trade must be £150,000 sterling. If his Majesty think fit to take the whole wool trade into his own hands for the good of his subjects of both Kingdoms and will advance the necessary stock, and will employ such fit commissioners as shall be humbly proposed to him, they will give sufficient security for the discharge of their trust, and not one stone of wool shall be transported into any country but England, and, at every year's end, all charges defrayed they will pay into the Irish Exchequer £30,000 sterling net besides the groats and usual fees on every stone to the Lord Lieutenant, provided that, if any war happen, they be allowed a sufficient convoy for the wool fleet.

But, if his Majesty does not do so, others of England and Ireland will advance the same, if commissionated thereto, no others to be licensed to transport any wool but those Undertakers for 21 years from 1 May, 1674. They will give the said rates for the wool, and if any difference arise in distinguishing the sorts, sworn officers, skilful in wool, may be appointed in every port, whose sentence shall be definitive, both to buyer and seller. They will enter sufficient security that not one stone shall be transported into any country but England, will pay the Lord Lieutenant his grace, and will pay yearly into the Irish Exchequer £10,000 sterling, provided they are allowed a safe convoy in time of war. The Commissioners or Undertakers will take over any contracts already made for the wool now shortly to be shorn.

The following advantages will arise from either of these proposals :—

1. No wool will be transported into foreign parts whereby their woollen manufactories will be discouraged and those of England flourish, and the Eastland trade and that with Turkey, Spain, Portugal and Muscovy will be regained, the French wool being too coarse and the Spanish too fine a mixture to English or Irish (and they will not mix with each other).

2. The Irish proprietors and tenants who are shipmasters will be encouraged and able to receive and pay their rents, which will also be no small advantage to the Crown.

3. The English wool will bear the higher rates, for, if the Commissioners or Undertakers are obliged to give such good rates, they must sell in England proportionably and that will keep up the price, whereas the Irish merchants by combination buying exceeding cheap in Ireland, can and do undersell the English wool sellers, so that they both in England and Ireland will be ruined.

4. It seems much for the benefit of England for the Irish wool to bear a good price when to be transported for England, for that will hinder its being manufactured in Ireland, which many years experience has showed nothing can induce them to, but mere interest.

Lastly, the Chief Governor of Ireland loses nothing of his right at 4d. a stone for licencing, and his Majesty gets an annual revenue of £30,000 if he advance the stock, and £10,000 if he does not, by what never yielded him anything and what better secures the effectual payment of his revenue in Ireland, and pleases both his English and his Irish subjects, and hinders foreigners from the possibility of growing in the woollen manufactures.

This cannot be judged a monopoly, since the propositions do not design the wool sellers should be obliged to it, but, if they will sell they bind themselves to give greater rates than have been given, and, since a monopoly is a restraint on trade in law, the licence is to take off the restraint laid on by law as to the persons to whom the grant is made, which is in favour of trade, and to an act of favour all people are not entitled, and therefore it may be given to one and not to another without injustice, for his Majesty is not obliged to give licence to any contrary to an Act of Parliament who, by his prerogative may dispense with that Act, without dispensations being for the good of his subjects, and he may choose whom he thinks fit to distribute his favour to.

APPENDIX XI.

AGREEMENTS IN CONNECTION WITH THE WOOLLEN MANUFACTURE AT CALLAN.

Articles of Agreement betwixt George Mathew, Esq., on the behalf of His Grace, the Duke of Ormond, etc., on the one part, and William Middleton, of Dublin, Clothier, on the other part, as followeth:—

That the said George Mathew, Esq., shall furnish the said William Middleton, with a convenient dwelling-house and other conveniences for work-houses in the town of Callan, at the rates expressed in his Grace's propositions for the planting of the said town, etc.

That the said George Mathew, Esq., shall supply the said William Middleton with wools such as he shall have occasion for from time to time at the rate of 8s. per stone, not exceeding £50 worth at a time; the said Middleton to give security to be responsible for the said wools, and to perform the conditions of the next article.

That the said Middleton shall cause to be spun the said wools at the town of Callan and country adjacent, and shall there weave the same in broad cloth according to the length, breadth, size and weight agreed upon betwixt the said William Middleton and Richard Laurence of Chapelizod, Esq., which cloth the said Middleton is to send to Dublin from the loom, and there deliver the same at his own proper cost to the said Richard Laurence or his assigns; and upon the certificate of the said Richard Laurence that he had received the said cloth well and sufficiently performed according to agreement, the said George Mathew, Esq., will pay the said Middleton five pounds for each cloth.

The said George Mathew, Esq., doth covenant to take, and the said William Middleton doth covenant to deliver, for the first year £500 worth of the said cloth, for the second year £1,000 worth, and for the third year £1,500 worth, and no more nor less, unless by a new or further agreement betwixt them.

Articles between Col. Richard Laurence and William Middleton.

It is agreed betwixt Richard Laurence and William Middleton, that the said Middleton shall make a sort of cloth known by the name of Wiltshire pack cloths, which are to contain out of the loom betwixt 41 and 42 yards, and full three yards broad within the lyst; the lyst of a deep blue, two inches broad; the cloth to be weaved in a ten hundred reed, and made of good fine fleece wool.

That the said contract shall continue for three years in case both persons find mutual encouragement; but if the said Middleton shall not be able to perform according to contract, he shall have liberty after three months warning, and delivering cloth for the stock of wools in his hand, to remove or stay and employ himself for his best advantage. And on the other hand, if the said George Mathew shall be discouraged to proceed, he shall have liberty, after six months warning, to withdraw his stock, allowing to the said Middleton £20 towards the charge of his remove, and other damages he may suffer by the trade undertaking.

(Ormond MSS., vol. iii., p. 348).

INDEX.

A.

	PAGE
Abjuration oath imposed	104
Absentee grantees	32
—— landowners	136, 167, 207
—— officials	203
Absenteeism alluded to	32, 235
—— causes insecurity of tenure	133
—— increase of	135
—— settlements designed to remedy	33
Absentees' losses through high rates of exchange	167
—— statute of	21
—— taxed	32, 33
Acts of Parliament in general, see Statues.	
—— in particular, see under subject (Cattle Acts, etc.).	
Additional duties, the	232
Adventurers, London citizens as	130
—— profligate of timber	147
Agrarian outrages, Act to suppress	219
Agriculture decayed	5
—— during revolutionary wars	104
—— state of	36-41, 142-144
Ale, see Beer.	
Ale-houses, Act to regulate	15
—— in Dublin	196
Alienation fines	134
Alnage-duty described	199
Alnage monopoly retained	61
—— office established	182
—— office, Dublin petition against an	182
Almoner, an, appointed	93
Alum imported	139
—— works at Kerry	149
Ambergris exported from Galway	82
American colonies compared with Ireland	224
—— emigration of sectaries to	155
—— export interest of	225
—— export of wool to	183
—— free trade with	173
—— trade with	117, 152
—— transportations to	102, 214
Anglesea (Lord) on Irish distress	155
Aqua-vitae, additional duties on	233

	PAGE
Aqua-vitae, barley grown for	41
—— monopoly of	15, 61
Arabia referred to	143
Arable land converted to pasture	43
Ardglyn iron works	49
Area of English rule	33
Arigna smelting works	49
Arklow glass works	82
Armagh linen manufactures	232
—— plantation of	23
Arms and ammunition, duty on	198
—— procured from abroad	95
Armstrong (Sir T.): his patent to coin copper	204, 208
Army, the, augmented	232
—— maintenance of	206
—— payment of	95
—— reduced	91
Artizans from abroad encouraged	151
Auction of confiscated lands	218
Avaux (Count of) on export of wool to France	178

B.

	PAGE
Bachelors, taxation of	233
Ballynegery glass works	83
Bandits numerous	104
Bandle, a, of cloth	184
Bandon, linen manufactures	78
Bankers, all brokers	210
Bank, proposed land	174
Bankers, ruined in 1680	210
Bantry, fishing at	193
Barbadoes, sugar trade with	207
—— transportations to	102, 103
Bark for tanning scarce	84
Barking of live oak trees	147
Barley for aqua vitae only	41
—— for beer	84
—— inferior	143
Barracks, cost of, met by tobacco tax	234
Barrel staves exported to France	147
Ballynegery glass works	83
Ballonakill iron works	52
Ballyregan iron mine	49
Beds and bedding of rushes and straw	141
Beef dear in England	165
—— exported to Holland	162, 165

INDEX.

	PAGE
Beef, price in Dublin in 1661	144
Beer as winter drink	139
—— consumption of	83, 84, 140
—— Excise on	196, 198
—— Excise on, raised	233
—— exported in 1685	196
—— plentiful	196
—— of poor quality	84
—— sale of, proclaimed	144
Beggars, Irish, in France	12
—— Irish, in London	12
—— numerous in Dublin in 1661	109
Begging uncommon	14
Belfast, forges near	49
—— pottery at, in 1688	195
—— shipbuilding at	196
—— shipping in 1663	169, 171
—— sugar refining at	197
Benzie (G. B.): his petition concerning woollen manufactures	183
Bigs (Sir A.): glass works at Birr	83
Birr, glass works at	83
Blackwater valley described	49
Blaney (Capt.): his account of the country cited	3
Bleaching yards, establishment of	231
Blenerhassett (Sir Leonard): his iron-works near Loch Erne	52
Blune (Charles), Lord Mountjoy and Earl of Devonshire: enters Cork	4
Blunt (Charles), &c., overthrows Earl of Tyrone	45
Boate (Gerard) on drainage of bogs	143
Boate (Gerard) on Irish cattle	40
—— —— on glass and pottery industry	83
—— on iron-works	106
—— on woods	44-47
Bog iron described	50, 51
Bogs, eggs and butter stored in	139
—— reclaimed, &c.	106, 143
Book of Rates raised	65
—— revised	62
Books of "Survey and Distribution"	28
Book of Transplanters' certificates	109
Boyle (Lord Chan.) on linen industry	191, 192
Brandy, importation of	144
Brehon laws	18
—— their nature	29
—— system	16
—— tenure, replaced by Knight-tenure	38
Brereton on cabins	36
—— on tillage	37
—— on land-purchase	40
Brewers petition against excise	196
—— of Dublin, obtain a charter	195
Brewing industry	83, 84, 195, 196
—— proposed tax on	84
Bridges scarce	68

	PAGE
Bristol merchants transport Irish as slaves	102
Brogues described	138, 140
Brogue-makers, timber spoilt by	147
Brook (Sir B.): his silver-mine at Kilmore	55
Brough, cattle destroyed at	158
Bullion, see coin.	
Bunrattie, Earl of Thomond's house there	54
Burnes, Irish Shipping at	163
Burning in the Straw	39, 143
Bushe (G. P.) on population	123
Butt (I.) on state of Ireland	238
Butter, export of	161, 162
—— stored in bogs	139

C.

Cabins described	137, 141
Cabins of Boughs, &c.	36
Caddowes (coarse blankets), manufacture of	72
Callan woollen factory	185
(App. XI.)	268, 269
Canary islands, English merchants at	163
Cannon cast at Cappoquin	49
Capital hard to raise	169
Cappoquin, cannon cast at	49
—— iron mine	49
Carbery, lands in possession of natives	20, 21
Carlow, co., well wooded	47
Carpenters numerous	142
Carrickfergus, K. William's arrival at	146
—— a staple established	72
Carrick linen factory, established by Ormond	190
Carrick-on-Suir, proposal to start brewery there	196
Carte on Elizabethan monetary policy	94
—— on prosperity of Ireland	11
—— on the plantations	38
Carteret (Sir Geo.) sends weavers from Jersey	189
Cary (John) on Cattle Acts	166, 227
—— on woollen and cattle trade	226
Catholic disabilities	13, 15, 124, 125, 216
—— hostility to woollen manufacture	221
Catholics, emigration of	214
—— excluded from teaching profession	13
—— impoverished	220
—— persecution of	214-216
Cattle, cottiers dependant on	137
—— destruction of	105, 212
—— foreign trade in	167
—— export abroad prohibited	61, 62
—— export to England prohibited	118, 154, 155, 207, 225
—— imported from Wales in 1652	105
—— imported in 1692	212
—— plague rampant	119
—— poor	40
—— scarcity of	212, 213
—— a staple commodity	154

INDEX.

Cattle Act, Ormond's protest
against (App. IV). - 246-8
—— Acts - - 157-170, 179-181
—— —— account of damage
inflicted by (App. V.) 248-254
—— —— destructive to trade 57
—— —— effect on tillage - 146
—— —— encourage sheep-
rearing - - 144, 146
—— —— memorandum urging
their repeal (App.
VIII.) - - 257-260
—— repeal advocated by
Cary - - - - 227
—— —— results of - - - 119
—— —— Sir Geo. Rawdon on 153
—— —— Sir Wm. Temple on
(App. VII) - 256, 257
Cavan, plantation of - - 23
Chapelizod linen factory
189, 190, 192
—— tenements for weavers - 189
Charcoal industry - - - 45
Charles I. and Irish trade - 63
—— graces demanded from - 32
—— Irish silver presented to 11
Charles II. and increase of
Irish duties - - - 174
—— his instructions to Lord
Robartes - - - 151
—— opposes the Cattle Acts - 118
—— duties granted to, by
Parliament - - 197, 8
Charleville linen factory es-
tablished by Lord Orrery - 190
Charters granted to trading
corporations - - - - 151
Chester, Irish cattle landed
there - - - - - 158
—— linen exported to - - 77
Chichester (Lord) and sea-
port towns - - - 89
—— on woollen manufactures 71
—— protests against prohibitions, &c. - - - 61
Chiefs and Clans - - - 16
Chimneys few - - - - 137
—— recently introduced
(1680) - - - - 141
Church, see Cottiers.
Civilization, Irish compared
with Greek - - - - 33
Civil wars, depopulation
caused by - - - - 144, 173
Cities established by James I. 60
Clan and English systems - 34
—— lands and clans - - 17-20
—— system extinguished - 18
Clansmen, emigration of - 34
Clare, settlement of - - - 20
—— transplantation to - 128
Clarendon (Lord) on scandal
of pensions - - 203
—— —— on Irish revenue - 234
Clergy over numerous - - 123
Clonmel woollen factory - - 186
Cloth industry restrained 177, 178
—— manufacture of - - 142
—— restrictions on export of 69
—— trade discouraged by
Wentworth - - - 74
—— —— English - - - 165
—— —— injured - - - 181

Clothworkers invited into Ireland - - - - 68, 69
Clothes, free import and export - - - - - 65
—— homespun and home-
made - - - - 175
Clothing industry, decline of 70
Coal anthracite - - - 48, 49
—— importation prohibited - 121
—— mines - - - - - 47-49
—— —— passing of Act to encourage - - - 121
—— —— little worked - - 150
—— pits in Kilkenny - - 121
—— used for smelting - - 149
Codd (John) authorized to
transport wool - - - 177
Coinage and credit 94-99, 203-216
—— counterfeit and clipped - 114
—— debased - - - - 61
—— effect of the Rebellion on 114
—— exempt from import duty 198
—— import of English,
stopped - - - - 65
—— remitted from England - 113
—— scarcity of - - 89, 113, 174
Coleraine planters interested
in tanning - - - - 85
Collins (John) on victualling
of ships - - - - 164
—— on results of Cattle Acts 166-7
—— on value of cattle and
sheep exports - - - 155
Colonial theory applied to
Ireland - - - - 223-225
Commerce with Spain attacked 4
Commercia' restraints - - 5
Common lands, enclosure of,
English - - - - - 38
Commons, House of, in-
terested in woollen in-
dustry - - - - 227
—— —— (English) petition
against Irish wool-
len industry - - 228
—— —— hostile to the Crown 222
Commonwealth money circu-
lated after the Restoration 206
Commissioners for Ireland,
orders - - - - - 102
Committee of trade appointed,
1662 - - - - - 150
—— —— condemn al-
nage office - - 182
—— —— continued under
William III. - - 219
—— —— Report on linen
industry - - - 187
—— —— report re re-
straints on export
of raw wool - 178, 179
Communication between
towns, means of - 67, 68
Composition of Connacht and
Clare in 1658 - - 19, 20
Confiscated lands auction - 218
Confiscations after Cromwellian war - - - - 116
—— cause insecurity of
tenure - - - - 41
—— crown rents from - - 87
—— effect on peasant pro-
prietary - - - - 130

INDEX.

Confiscations following refusal of abjuration oath - - - - - 104
— of Catholic property 217, 218
— of land - - - - 15, 31
Connaught, cattle produced in 146
— collective ownership in - 28
— composition of - - - 87
— confiscation of - - - 25
— deportation of land owners to - - - 102
— small freeholders secure in - - - - - 30
— many woods in - - - 47
— settlement of 20, 26, 129-132
— sparsely populated - - 89
— transplantation to 102, 109, 128
Conny island, Sligo, silver mine at - - - - - 54
Construction, period of - 6, 8-99
Coote (Sir Chas.), iron works started by - - - - 52
Copenhagen palace, Irish woollens as hangings there 227
Copper coinage - - - - 97
— coins, patent for making 208
— mine at Kenmare - - 150
— mines unimportant - - 56
Cork city in 1647 - - - 112
— fisheries at - - - - 193
— — invaded by French - 194
— Mountjoy's entry into - 4
— plantation of, proposed - 133
— pewter manufacture at - 197
— shipping belonging to - 171
— woollen industry - - 186
— (Earl of) encourages fishing industry - - - 82
— — encourages iron works - - 49, 52
— — his silver and lead mines - - - 53
— — on condition of Ireland in 1630 - - 40
Corn burning in the straw - 143
— burnt by order of Lord Justices - - - - 105
— cultivation of - 41-43, 143
— export prohibited - - 58
— plentiful - - - - 146
— proclamation (English) against importation of 145
— scarcity of - - - 4, 212
— Table of exports in 1665 145
Corporations, rights of restrained - - - - - 151
"Coshering," practice of - 34
Cottages. see Cabins.
Cottier-class - - - 14, 16
Cottiers. the - - 33-36, 136-142
— little affected by Cromwellian Settlement - 117
Counterfeiting prevalent - 206
Council of trade - - - 151, 170
— — protest against Cattle Acts - - 154
Court of claims - - - - 207
— Wards and Liveries Instituted - - - 91
— — — abolished - - 202
— — — profits derived from - 87

Coyne and Livery, customs of 37
Craftsmen work at home - 176
Creaghting abolished - - 30
— described - - - - 29
— injurious to agriculture 37
Crommelin (Louis) settles at Lisburn - - - - - 232
— (Henry), petition to - - 111
Cromwell (O.), conquest by - 103
— taxation by, continued after Restoration - - 197
Crown, interest of, in Irish prosperity - - 9, 10, 223
— Rents - - - - - 197
— Revenues dependant on trade - - - - 86, 88
— title - - - - - 21
Currency, see Coinage.
Custom House Officers corrupt - - - - - 61
— — — oppressive - - 65
— misappropriated - - 89
"Custom of the Country," recognition of - - - 116
Customs and excise, Patriot parliament remit - 171, 172
— on cattle - - - - 166
— amount in 1661 - - - 110
— their nature - - - 198
— amount of (table) 1632-1640 - - - - - 64

D.

Dairy farming in England - 162
— farming, primitive - - 161
Davies (Sir John) on Irish labourers - - - - - 14
— on land tenures 18, 19, 21, 37, 38
— schemes for plantations - 33
— his settlement of tenures 8
Depopulation in 1582 - - 3
— in 1652 - - - - 101
— in 1656 - - - - 103
— following forfeitures - 133
Derry, see Londonderry.
Desert land iron-mines - - 51
Destruction, period of - 100-115
Desmond, settlement of - - 20
Desolation following the Revolution - - - - - 213
Devastation general in 1652-53 - - - - - - 108
Diet of common people, the 139
Donegal co., Forests in - - 47
— settlement of - - 21, 23
Douballie, Cavan, iron-works 51
Dress, native - - 70, 138, 140
Drogheda linen industry - 231
Drunkenness prevalent in 1600 - - - - - - 15
— uncommon - - - - 126
Dublin, ale-houses near, in 1676 - - - - - - 196
— beggars numerous in 1609 - - - - - - 102
— brewers obtain charter - - - 195, 196
— food scarce and dear - 5
— glass-works at - - - 195
— linen industry in 1691 - 231

INDEX.

Dublin mint, established by James II. - - - 209
— mint, proposed erection of - - - - - 114
— partially in ruins in 1645 112
— shipping - - - - 171
— rebuilding of - - - 147
— Table of prices there, 1599-1602 (Appendix I.) 241
— trade ruined - - - 110
— weavers, petition against the Alnage office - - 182
— woods scarce near - - 46
Duhallow, clanlands granted to chief at - - - - 20
Dunally mines wrecked - - 106
Dunbar (Sir John): his ironworks in Fermanagh - - 52
Dunboyne (Lord), possessions of - - - - - - 109
Dundalk, no woods near - 46
Dungannon (Lord) on Cattle Act - - - - - - 157
Dungarvan, fishing off - - 195
Dunkirk, trade with - - 163
Dutch and Irish shipping - 67
— doggers captured in 1672 194
— fishermen off Irish coast 81
— immigrants start woollen companies - - - 186
— privateers active - 167, 168
— shipping, hire of - - 170
— war injurious to Irish trade - - - 155, 163
— helps Irish shipping - 170
— wars: effect on Ireland 167 on
Duties, additional, imposed 232-233
— import and export - 59, 87
Dyes and dyeing - - - 139

E.

Earls, flight of the - - 12, 22
East India Co. iron works, &c., in Munster - - 85
— Spanish coin used by - 98
Elizabeth, coinage debased by - - 94, 95, 209, 210
— plantations under - 16, 33
— regrants by - - - - 19
— trade policy of - - - 5
Elizabethan wars, devastation after - - - 236
— — ravages of - - - 41
— — woods destroyed in 43
Emigration, cessation of, in 1688 - - - - 119
— continuous - - - 12
— general - - 210, 211, 219
— of freeholders - - - 25
— of non-conformists - - 155
— of Protestant woollen-workers - - - 221
Empire, birth of the British 224
Employers and wage earners 176
Employment in 1672, amount of - - - - - - 145
Enclosures few in number - 143
Engrossing coin practised - 98
Enniscorthy iron works - - 149
Exchange, rates of, high - 206
— — raised - - 207, 208

Excise, nature of the - - 198
— on beer revised - - - 233
— on tobacco - - - - 234
Exeter weavers destroy their looms - - - - - 185
Explanation, Act of - 124, 131, 132
Export, foreign - - 67, 160, 161
— of cattle to England prohibited - - - - 155
— — Scotland prohibited - - 153, 157
— of corn in 1641 - - - 43
— — prohibited - - 146
— of glass prohibited - - 120
— of horses to Scotland prohibited - - - 157
— of iron in 1641 - - - 49
— of linen restrained - - 154
— of linen yarn in 1641 - 80
— of manufactures, duties on - - - - - - 117
— of mutton, &c., to England prohibited - - 157
— of timber prohibited - 147
— of wool restrained - - 154
— of woollens prohibited - 120
— Table showing articles of 66
— to American plantations limited - - - - 152
Export trade hampered - - 175
Exports, removal of restraints on - - - 62, 178
Exports to England in 1614 - 42
— value of, in 1668 - - 170
— Wentworth advocates free - - - - - 10

F.

Factors, trade by English - 113
Fairs established by James I. 60
Famine, numbers perish from 212
— threatened in 1643 - - 106
Farmers of revenue, absentees - - - - - - 207
Felon's goods, forfeiture of - 87
Fermanagh, co., iron works 51, 52
— plantations of - - - 23
Finance, public, see Revenue.
Fines mentioned - - - 198
Fish, exports in 1641, table 82
— 1665 and 1669, table - 193
Fisheries, attempt to improve - - - - - 193-4
— exclusion of foreigners from - - 81 and 194
— foreigners gain profits of 80
— invaded by French fishermen - - - - - 194
— ruin of - - - - - 112
Fishermen, tax on foreign - 80
Fishing industry - 80-82, 193-195
— Letter from Kinsale on - 194-5
— trade encouraged by William III. - - - - 219
— practised - - - - 125
FitzWilliam, Lord Deputy, and export of wool 69, 70
Flax, cultivation of - - 77-79
— cultivation suspended - 137
— dressing of - - - - 142
— free imports in 1695 - 231

INDEX.

Flax seed, import of - - - 190
Fleet, Irishmen pressed for service in - - - - 112
—— victualling of - - - 150
Flight of the Earls - - 12, 22
Food, cost of - - - 5, 138
—— mainly milk and potatoes - - - - 141
—— production - - - - 173
Foreign currency, circulation of - - - - 204-5, 208-9
Foreigners encouraged to settle - - - - - 174
Forfeitures after Cromwellian war - - - - 127
—— extent of - - - - 132
—— Temple on - - - - 142
—— quit rents arising from - 198
France, emigrations to - - 214
—— England's struggle with - 223
—— political ally of Ireland 223
—— peasantry of - - - 138
—— wool smuggled into - 180, 181
—— woollen trade with - 177, 178
Freedom of cities, &c. - - 172
Freeholders, Catholics incapable of being - - - - 216
—— numerous - - - 34, 35
—— small, crushed out of existence - - - - 25
Free trade with Scotland and Isle of Man in 1652 - - 110
—— agitation - - - - 221
French fishermen invade Irish waters - - - - 194
Fuel, mainly turf - - - 139
Fuller's earth, import of, forbidden - - - 73, 185
Furniture, absence of - - 140
Furs, trade in - - - - 197

G.

Galleys, Scotch built, of Irish timber - - - - - 43
Galway city charter - - - 69
—— merchants export ambergris - - - - 82
Gavelkind abolished - - 22
—— still used - - - - 28
Giraldus Cambrensis - - 44
Germany, gold in - - - 57
Glankankin woods - - - 47
Glass, black - - - - 83
—— clay imported - - - 83
—— industry - - - 83, 195
—— monopoly - - - 60, 61
—— works - - - 83, 195
Gold mines in Kerry - - 53, 56
Gookin on husbandry - - 142
Gough (C.): his "Forfeiting Proprietors" - - - - 26
Graces, the, deficit met by - 91
—— the, demanded - - - 32
—— evaded - - - - - 62
—— granted - - - - 88
Grandison (Lord), founder of Merchant Staplers - - - 75
Grantees, absentee - - - 32
Granaries established - - 145
Grattan's Parliament revives woollen industry - - - 230

Grazing industry fostered - 235
"Green wax money" - - 87
Greyhounds - - - - 34
Grievances, articles of - - 65
Griddles used for baking - 141

H.

Hamilton, Sir Geo., owns silver mine at Kilmore 53-55, 148
—— —— —— - - - - 148
—— Lady Mary: her poverty 109
Harding, W. H., on number of confiscated estates 26
—— on prosperity of peasant class - - - - 130, 136
Hart (Sir Percival): monopoly to make black glass - - 83
Harvest bad in 1674 and 1687 119
Head-dresses, linen, worn by women - - - - - 232
Hearth-money - - - 198, 199
—— oppressive - - - - 202
—— returns - - - - 123
Hemp, free imports in 1695 - 231
—— import duties on - - 198
—— sowing, ceases - - - 137
—— sowing enforced - - 188
—— women skilful in dressing 142
Henry II.'s conquest - 44, 45, 101
—— VIII., land schemes of - 21
Hewson (Coll.) at Wicklow - 106
Hides, exports in 1608 - - 84
—— exports in 1665, 1669, and 1685 - - - - 197
—— small and poor - - - 161
—— badly prepared - - - 162
—— trade with France in 1691 - - - - - 164
—— English trade unprofitable - - - - - 166
Holinshed on Ireland - - 3
Holland, beef exported to - 165
—— brewers from - - - 196
—— flax seed imported from 79
—— rates of interest in - - 168
—— trade with - - - - 164
—— wool export to, prohibited - - - - 177
—— wool smuggled into - 181
—— see also Dutch.
Holy Days regulated by statute - - - - - 126
Homage, respite of - - - 87
Homespun clothes - - - 141
—— frieze generally worn - 184
Hops, imports of, in 1685 - 196
Horses, account of - - - 144
—— destruction of - - - 105
—— Irish hobby - - - - 40
—— import of - - 41, 63-65
—— seized by soldiers - - 212
Hospitality, general - - - 140
Hounslow Heath, Irish encamped on - - - - 222
Houses of Correction established - - - - - 14
House property destroyed - 106
Housing, account of - - - 138
Huguenots establish woollen manufactures - - - - 186
Husbandry, see Agriculture.

INDEX.

I.

	PAGE
Idleness of dispossessed gentry	14
Idough, coal mine at	150
—— coal and iron mines near	48
Ikerrin (Visct.): very poor	109
Immigrations, statistics of	122
—— Protestant, from France and Netherlands	123
Import of cloth, etc., from England	70
—— of English goods, cessation of	164
Imports exceed exports in 1641	66
Inchiquin, Lord: action at Cork	112
—— cessation with	106-7
Indigo, import of	139
Industrial Revolution, The	175
Industries, attack on	219
—— in general	173-176
—— (minor)	83-85
—— ——	195-197
Industry, destruction of	5
—— general	14
Inns, lack of	68
—— Act to regulate alehouses, &c.	15
Intermarriages between English and Irish	134
Interest, high rate of	97, 99, 168, 169
Intolerance, outburst of	215
Iron, export to England	224
—— exports and imports compared	149
—— imported from Spain and Sweden	149
—— works	48-53, 149
—— —— destroyed	53
—— —— effect on timber supply	44, 46, 147
Isle of Man, free trade with	110
Italy, linens exported to	77
—— woollens exported to	183

J.

James I. gives Charters to maritime towns	58
—— on Irish ploughs	36
—— reign of	8
—— II., currency debased by	209-210
—— mints established by	209
Jersey weavers, immigration of	189
Jewels exempt from duty	198
Justice, difficulty of executing	133

K.

Kenmare copper mine	150
Kern, see Tory.	
Kerry alum works	149
—— gold and silver mines	53
—— plantation of	133
Kilkenny Confederates issue coins	114
Killulta, woods in	47
Kilmacoe iron mine	49
Kilmore silver mine	54, 55
Kilwarlin, woods in	47

	PAGE
King (W.), Archbp., on attacks on Irish trade	220
King's Co., woods in	47
Kinsale fisheries	193-195
Knight's Service replaced by socage	134
—— —— revived	90
Knox (Sir J.): his patent to coin copper	208

L.

Labour, division of	176
Labourers badly paid	14
—— improved condition of	136
—— scarce	103, 133
Land grabbers	131
Land owners dispossessed	237
—— —— penalized	121
Land, registries of, suggested	150
Landlord and tenant	176
Land-system, the	8, 127-136
—— —— changes in the	216
—— —— criticized	235
—— —— settlement of the	15-33
Land-tax imposed	233, 234
Land-tenures, fixity of	29
—— —— in capite revived	90
—— —— insecure	41, 44, 132, 133, 142, 147
—— —— Settlement of, by Davies	8
—— —— unsatisfactory	37
Land-values depressed	168
—— —— raised	11, 38, 40, 119
La Rochelle, Siege of	82
—— Trade with	163
Lawrence (Col. R.): his linen factory at Chapelizod	190, 192
—— on absentee officials	203
—— on Callen woollen works	185, 186
—— on Cattle Act	171
—— on Council of trade	151
—— on idlers	123
—— on national character	133
Lead exported	53
—— mines	53-56
Lead-ore	149
Leases, account of	27-30
—— nature of	134, 135
—— too short	37
Legge (George): his patent to coin copper	208
Legislative independence affirmed	120, 173
—— —— attacked	223, 225
—— —— destroyed	238
Leinster lands claimed by Crown	21
—— plantation of	28, 33
—— woods in	47
Leitrim, co., ironworks	49, 52
—— plantation of	24, 25, 29, 78
Lent, revival of	155
License duties	198
Licensees, fees for	87
—— to export prohibited goods	58
—— to export wool	179, 180
Lighthouse duties	197, 199
Limerick mint established	209

INDEX.

Limerick plantation - - 27, 133
—— price of lead at - - - 55
—— siege of - - - - 214
—— treaty of, violated - - 215
—— woollen industry - - 183
Linen Co. fails - - - - 231
—— exported free of duty - 111
—— exports and imports—
  table - - - - 189
—— head dresses worn - - 232
—— industry 75-80, 185-192, 230-232
—— industry develops - - 119
—— manufacture of - - - 142
—— manufacture encouraged
  150, 174, 185, 226
—— trade encouraged by
  William III. - - - 219
—— yarn seized - - - - 65
Liquor traffic - - - - 126
Lisburn, Crommelin's settlement at - - - - - 232
Lisfinnon iron mine - - - 49
Lishan river iron works - - 51
Lithgow on tillage - - 36, 37
Living, cost of, cheap - - 140
—— cost of, increased - - 95
Locke (John) on Armagh
  diaper - - - - - 232
Loftus (Sir A.), Visct. Ely: his
  ironworks - - - - 52
London Co., the, tenants - 27, 28
Londonderry (Earl of): his
  ironworks - - - 52
—— plantation - - - - 23
—— planters encourage tan-
  ning - - - - 85
Longford, plantation - 24, 25
Lough Allen, ironworks near 51
—— Erne, woods near - - 47
—— —— iron works - - 51, 52
—— Neagh ironworks - - 51
Lovett (C.) granted Chapel-
  izod bleach yards, &c. - 192
Low-Countries, linen industry
  in the - - - - 189

M.

MacCarthy of Muskerry,
  grant to - - - - - 19
McGuinness of Iveagh (Sir
  A.), lands of - - - - 19
McMahon's country regranted 19
Madder imported - - - 139
Maguire (Connor Roe), land
  granted to - - - - 23
Manufactures, English im-
  port duties on - - - 117
—— export of prohibited - 225
—— small demand for - - 173
Manure, fish used as - - 195
—— various kinds of - 39, 40
Maritime towns granted new
  charters - - - - 59
Markets, foreign and English 174
Masons numerous - - - 142
Mayo, shipbuilding at - - 85
Merchants, credit of - - 168
—— condition of - - 168, 169
—— foreign and English
  discouraged - - - 61
Metal exports, table of - - 148
Milch kine, blowing of - - 39

Mills destroyed - - - - 105
Minehead, silver and lead
  mine - - - - - - 53
Mines - - - 47-57, 148-150
Mint, establishment of,
  discussed - - 99, 115
—— —— —— proposed - - 205
Mints established - - - 209
Mocollop iron mine - - - 49
Molyneux defends legislative
  independence - - 225
—— on linen industry - - 231
Money, see Coinage and Revenue.
Monopolies, various - - 60, 61, 65
Monopoly of selling wine, &c. 15
—— pottery - - - - - 195
—— salt, proposed (App. III.)
  244, 245
Mountrath ironworks 51, 52, 149
Morris (Dame Katherine):
  her poverty - - - - 109
Moryson (Fynes) on seizures
  of land - - - - - 4
Mountjoy (Lord), see Blunt
  (Sir Chas.).
Mountmellick, ironworks
  near - - - - 51, 52
Munster, fishing industry encouraged - - - - 82
—— glass industry - - - 83
—— ironworks - - - 49, 52, 85
—— merchant adventurers receive charter - - - 151
—— plantation of - 16, 20, 26, 27
—— president of: see Tho-
  mond (D.), Earl of.
—— protest against staples - 72
—— shipbuilding - - - 85
—— woods destroyed - 43, 47
—— woollen manufactures - 73
Murano, manufacture of
  black glass - - - - 83
Murrain, frequent outbreak
  of - - - - - - 40

N.

Naas: its favourable position 104
—— untilled in 1643 - - - 105
Naturalisation of Protestants 174
Navigation Acts impede trade
  with America - - 57, 174
—— —— provisions of - 111-112
—— —— restrictions under - 67
—— —— results of - - 152, 153
—— —— stop export of woollens - - - - - 178
—— —— unjust - - - - 117
Netherlands, linen -exported
  to the - - - - - - 77
Newry, no woods near - - 46
Non-importation agreements 159

O.

Oates', Titus, conspiracy - 124
Oats, see corn.
O'Byrne Clan, migration of
  the - - - - - - 18
O'Doherty's granary burnt - 41
O'Donnell Clan, the regrant
  to - - - - - - - 19
O'Dowds of Sligo, clanlands - 18

INDEX.

O'Ferralls of Longford, clanlands - - - - - 18
Oil mills - - - - - 197
O'Kellys of Galway, clanlands 18
O'Kennedy (John MacDermot): his silver mine - - 54
—— (Hugh) destroys Kilmore silver mine - - - 55
O'Neile (Maj. Harry), licensed to transport wool to France, &c. - - - - 177
O'Neill, The, receives regrant 19
—— (Con): his curse - - - 41
Ordnance manufacture contemplated - - - - 49
Ormond (Duke of) advocates a public Bank - - 210
—— —— and fishing industry 194
—— —— and wool industry 178, 185
—— —— encourages Irish industries - - - - 174
—— —— encourages trade - 150
—— —— English jealousy of - 118
—— —— fosters linen industry - - - 189, 190
—— —— his protest against Cattle Act (App. IV.) - - - 246-248
—— —— on absenteeism - 135, 136
—— —— on cattle export - 156
—— —— on Dutch war - 167-168
—— —— on excise on beer and ale - - - - 196
—— —— on Linen Act - - 188
—— —— on oppression of Ireland - - - - 152
—— —— on trade - - - 125
—— —— stops export of live cattle, &c. - - - 118
Orrery (Lord) establishes linen factory at Charleville 190
Ostend, trade with - - - 163
O'Toole, The, clanlands - - 18

P.

Pale, much corn grown in the 42
—— security in the - - - 104
—— want in the - - - - 5
Parliament, Catholics excluded from - - - - 215
—— English hostile to Ireland - - - - 118, 237
—— English, its supremacy - 237
—— English, legislates for Ireland - - - - 222, 223
—— Patriot, claims legislative independence 120, 173
—— —— land legislation proposed by - - - 216
—— —— proposals of the 120-122
—— —— trade act of - 171-173 (App. IX.) - - 261-264
Pasture and tillage 41-43, 144-146
Peasant proprietors - 17, 30, 130
Peasants, condition of - - 137
Pedlars exempted from Vagrancy Acts - - - 67
Penal laws alluded to, 6, 13, 33, 104, 119, 128, 215, 235
—— see also Catholic disabilities.
Pensions very numerous - 203

Penzance, Irish vessel wrecked at - - - - 163
" Perus," circulation of - - 114
Petitions, English hostile - 225
Pett (Sir Peter): his memorial 183
Petty (Sir Wm.): his coppermine at Kenmare - 150
—— —— his estimate of population - 12, 123
—— —— on Absenteeism - 135
—— —— on Alum works - - 149
—— —— on damage caused by rebellion - 109
—— —— on Dublin ale-houses and breweries - 196
—— —— on farming of revenue - - - 202
—— —— on finances - - 207
—— —— on fishing industry - 193
—— —— on housing, etc. - - 138
—— —— on iron works - - 149
—— —— on land values 132, 133
—— —— on laziness - - - 125
—— —— on Linen Act - - 188
—— —— on number of chimneys - - - 175
—— —— on sheep and wool - 184
—— —— on shipping - - 170
—— —— on tanning industry 197
—— —— on wealth of Ireland 150
Pewter, manufacture of - - 197
Piggot (Sergt.-Maj.): his iron mine - - - - - 51
" Pins," a form of iron mine 52
Pinnar ( ) on rents in Ulster - - - - - 27
Pirates dispersed by Wentworth - - 64, 82
—— infest South coast - - 63
—— numerous - - - - 110
—— trade with - - - - 58
Plague follows revolution - 212
—— ravages by - - - - 155
Planters encourage tanning - 85
—— mainly soldiers or citizens - - - - - 142
—— Sir Wm. Temple's account of the - - - 103
Plantation lands - - - - 40
Plantations - - - - - 20-29
—— Carte on the - - - 38
—— their importance - - 15
Ploughing by the tail 38, 39, 142
Pliny referred to - - - 57
Poll-tax imposed - - - 233
—— return of - - - - 123
—— table showing amounts levied - - - - 233
Population, Petty's estimate of - - - - 12, 122
—— South's estimate in 1696 123
—— in 1600 - - - - - 12
—— self-supporting - - - 175
Portarlington, glass works at 195
Port-dues exorbitant - - - 59
Potatoes, introduction of - 14
—— a staple aricle of food 107, 137, 139, 140, 146
Pottery industry - - 83, 195
Poundage, its nature - - 198
Poyning's Act referred to - 65
Privateers, Dutch, losses by 167, 168

INDEX.

Prizage described - - 199
—— imposed - - - - 197
Proclamation, attempt to raise taxes by - - - 88
—— on prosperity of Ireland 11
Produce, Irish, excluded from English market - - - 117
Production, primitive methods of - - - 175, 176
Professions closed to Catholics - - - - - 216
Property, destruction of 2, 212, 213
Prosperity of Ireland in 1632 11
Provisions abundant - - 126
—— export of, forbidden - 111
—— exported - - - - 163
—— exports of in 1665 and 1669 - - - - - 161
—— imported from England in 1603 - - - - 5
—— trade in - - - 118, 161
Provision trade (Sir Wm. Temple's proposals on the) (App. VI.) - - 254-256
Protestant nationalism, beginnings of - - - - 220
Public works, execution of - 94

Q.

Quakers, losses incurred by, during revolution - - 213
Queen's Co., woods in - - 47
—— mines in - - - - 52
"Quia Emptores," statute of 17
Quicksilver found at Kilmore mine - - - - - - 55
Quit rents described - - - 198
—— —— loan secured on - - 232

R.

Ranelagh (Lord) farms revenue - - - - 201, 202
Rapparees, see Tories.
Rape-oil mills - - - - 197
Rawdon (Sir Geo.) assists fisheries - - - 194
—— —— on alnage office - 182
—— —— on Navigation Act - 153
Rebellion in 1611 - - 6, 31, 66
—— of 1641, consequences of 101-150
—— —— forfeitures following 198
—— —— land tenure previous to - - - - 25
—— devastations of - - - 154
—— ravages during the - 109, 236
Rebellions frequent - - - 132
Reconstruction, period of 6, 116-210
Recuperation, power of 11, 119, 214, 237
Redistrbution, period of 7, 211-238
Regrant of clanland, policy of - - - - - -17-20
Religious disabilities - - 135
Rent of English and Irish lands compared - - - 165
—— low in England - - 155
Rents, Rack, prevalent - - 235
Restoration, absenteeism revived by the - - - - 33

Restoration, metal exports after the - - - - 148
—— revenue at the - - 114, 202
—— tillage at the - - - 144
—— trade restrictions following the - - - - 57
Restraints on exports removed 159
Revenue - - 85-94, 197-203
—— deficit after the revolution - - - - - 232
—— depressed - - - 108, 234
—— during rebellion - 131, 132
—— in 1649-56, abstract of - 113
—— inadequate - - - - 113
—— increased - - - - 119
—— tables of (1628) 91, (1640) 93, (1663-4) 200, (1683-5) 200
—— under the Stuarts - - 58
Revolution, desolate state of Ireland after the - - 237
—— disastrous to Ireland - 120
—— outbreak of - - - - 211
—— results of - - - - 211
Rhaninus berries - - - 139
Rivers, condition of, in 1662 68
Roads in 1662, building of - 68
Robartes (Lord), Charles I.'s instructions to - - - 151
Rock, Capt., see Tories.
Rome, appeals to, prohibited 13
Roper (Sir T.): cloth manufactory in Dublin - - 72
Roman Catholic, see Catholic.
Russell (Sir Wm.), proprietor of silver mine at Kilmore - 55

S.

St. John (Sir O.), on clans - 17
—— on exports to England - 42
St. Leger (Sir W.) imports corn - - - - - - 42
Salt, attempted monopoly of 60
—— proposed monopoly of 10, 244, 245
—— scarce - - - - 82, 195
—— tax on - - - - - 193
Scariff, iron works at - - 49
Schools of navigation, &c., started - - - - - 172
Scotch settlers and linen industry - - - 231
—— —— industrious - - 42
—— —— trespass on fisheries 194
Scotland, free trade with - 110
—— import of wool from, forbidden - - - - 182
—— troubles in, 1679 - - 168
Sea coal - - - - - 46
Serfdom, condition of - 34, 35
Settlement, Act of - 147, 198, 217
—— (Cromwellian) - 35, 100, 113
—— —— and agriculture - 142
—— —— causes absenteeism - 135
—— —— results of the - 127, 129
Settlements (Restoration) - 131
—— —— cause absenteeism - 135
—— (Tudor and early Stuart) 53
Seventeenth and eighteenth centuries compared - - 1
Shane (Sir Fras.) on state of Ireland - - - - - 3
Shannon, coal near the - - 47

INDEX.

Shannon, Wentworth's plan to utilize the - - - 68
Sheep described - - - 141
—— destroyed - - - 212, 221
—— displace cattle 118, 144, 146, 162, 164, 165
—— numerous - - - - 184
—— owners - - - - 179
—— pulling of - - - - 39
—— rearing - - - - 72
—— slaughter of, proclaimed 176
—— see also Wool and Woollen industry.
Sheridan (T.): his proposal for buying all wool (Appendix X.) - - - - 265-268
Shipbuilding - - 85, 148, 170
—— —— at Belfast - - 196
Shipping, account of 67, 111, 169-173
—— increased - - - 150, 164
—— ruined - - - - - 169
—— (Dutch) hired - - - 170
—— (English) victual Ireland 165
Sidney (Lord Deputy) restricts wool exports - - 69
—— (Lord) on trade - - - 219
Silver and silver mines - - 53-56
—— patent to coin - - - 205
—— see also Coinage.
Skenes, manufacture of - - 49
Slavery introduced - 102, 103, 214
Slewgalen, mt., iron works near - - - - - - 51
Slew Neron iron works - - 51
Smith (Col. Thos.) on churls - 34
Smoking universally practised - - - - - - 140
Smuggling general - - 58, 180
—— of cattle and horses - - 158
Snuff much used - - - 140
Socage replaces Knight's tenure - - - - - 134
Soldiers emigrate - - - 102
—— used in England - - 222
South (Capt.): his estimate of population in 1696 - - 123
Spain, ambergrease exported to - - - - - 82
—— corn exported to - - 41
—— fish exported to - - - 195
—— glass clay imported from - - - - - 83
—— soldiers emigrate - - 102
—— trade with - - - 59, 164
—— wool exported to - - 183
—— wool smuggled to - - 180
Spanish auxiliaries - - - 95
—— coins circulated - - 97, 98
—— fleet, victualling of the - 63
Spencer (E.) on the plantations - - - - - 27
Spirits, excise on - - - 198
Staple-towns founded - 71, 72
Star-Chamber fines - - - 87
Starvation general - - - 108
Statutes cited:—(2 Anne, c. 6) 217, 218; (8 Anne, c. 6) 218; (10 Car. I., c. 1) 91; (10 Car. I., c. 2) 91; (10 Car. I., c. 23) 91, 99; (10 Car. I., sess. 3, c. 26) 68; (10 Car. I., sess. 3, c. 31) 34; (10 & 11 Car. I., c. 5) 15; (10 & 11 Car. I., c, 15) 14, 39; (10 & 11 Car. I., c. 16) 34; (10 & 11 Car. I., c. 17) 39; 15 Car. I., c. 13) 94; (12 Car. II., c. 4) 178; (12 Car. II., c. 18) 152; (13 Car. II., c. 23) 157; (14 Car. II., sess. 2, c. 2) 198; (14 & 15 Car. II., c. 6) 198, 199; (14 & 15 Car. II., c. 7) 199; (14 & 15 Car. II., c. 8) 196, 198; (14 & 15 Car. II., c. 9) 182, 198; (14 & 15 Car. II., c. 13) 174; (14 & 15 Car. II., c. 17) 199 (14 & 15 Car. II., c. 18) 126 (15 Car. II., c. 7) 152; (20 Car. II., c. 7) 157; (17 & 18 Car. II., c. 8, sec. 36) 124; (17 & 18 Car. II., c. 9) 137, 188; (17 & 18 Car. II., c. 15) 182, 199; (17 & 18 Car. II., c. 17) 199; (17 & 18 Car. II., c. 18) 199; (17 & 18 Car. II., c. 19) 126, 198; (20 Car. II., c. 7) 157; (22 & 23 Car. II., c. 2) 157; (22 & 23 Car. II., c. 26) 152; (32 Car. II., c. 2) 157; (5 Ed. IV., c. 6) 80; (12 Ed. IV., prohibiting exports) 58; (2 Eliz., c. 3) 13; (11 Eliz., sess. 3, c. 10) 58, 69; (11 Eliz., c. 3) 84; (11 Eliz., c. 5) 77; (13 Eliz., c. 1 & c. 2) 69; (19 & 20 Geo. III., c. 20) 182 (13 Hen. VIII., c. 2) 69; (28 Hen. VIII., caps. 5, 6, & 13) 13; (33 Hen. VIII., c. 2) 77; (10, 11, & 12 Jac. I., c. 10) 90; (11, 12, and 13 Jac. I., c. 5) 59; (11, 12, & 13 Jac. I., c. 7) 68; (Will. & Mary, c. 3) 233; (7 Will. III., c. 12) 218; (7 Will. III., c. 14) 126; (7 Will. III., c. 15) 233; (7 Will. III., c. 21) 219; (7 & 8 Will. III., c. 39, Eng.) 231; (9 Will. III., c. 8) 233; (9 Will. III., c. 14) 197; (10 Will. III., c. 3) 233; (10 Will. III., c. 12) 148; (10 Will. III., caps. 4 & 5) 234; (10 Will. III., c. 5) 229; (10 & 11 Will. III., c. 10) 229.
Stevens (John) on diet, &c., of natives - - - - 140
Strabo referred to - - - 57
Strafford (Earl), see under Wentworth (Sir T.).
Stuart kings befriend Ireland 237
—— —— policy in Ireland - 221
Stuarts' struggles with the Parliament - - - 58, 86
Subsidies described - - - 90-93
—— grant of - - - 88, 199, 200
—— paid in corn - 145, 200, 201
—— table of - - - - - 92
—— unpaid - - - - - 155
Sugar plantations, transportations to the - - - 102
—— refining introduced - 196, 197
—— trade with Barbadoes - 207

INDEX.

Surrender and regrant, policy of - - - - 16
Swift (Dean): his political writings - - - - 220

T.

Tables:—(Corn exports) 43 & 145; (Customs collected) 64; (Customs duties) 62; (Damage caused by the Rebellion) 109; (Exports in general) 66; (Fees on wool) 179; (Fish exports) 89, 193; (Foreign & English coin values) 209; (Foreign Coin values) 205, 208; (Linen exports and imports) 189; (Metal exports) 148; (Poll-tax returns) 233; (Population) 122; (Prices in Dublin market) 241; (Provision exports) 161; (Revenue) 91, 93, 113, 200; (Seeds & Crops per acre) 143; (Subsidies) 92; (Timber exports) 43; (Wool exports) 75, 183; (Woollen exports) 186, 227; (Woollen imports) - - 228
Tallow (co. Cork) iron mine - 51
Tallow, account of - - - 161
—— English trade in - - 166
—— restraints on - - - 64
Tanistry abolished - - - 22
Tanning industry - 84, 85, 197
Tariff, English, re-drafted - 152
Taverns, brewing in - - - 196
—— see also Ale-houses and Inns.
Taxation, excessive - - - 202
—— hinders tillage - - - 41
—— increased - - - 232, 234
—— struggle concerning - 86
Taxes, proposal to pay, in flax and hemp - - - 191
Temple (Sir Wm.) on absenteeism - - - - 135
—— on Cattle Acts - - 256, 257
—— on cloth manufactures 184, 185
—— on English settlers - - 103
—— on fishing industry - - 193
—— on horses and sheep - 144
—— on land values - - - 119
—— on linen industry - 190, 191
—— on ploughing by the tail 142
—— on provision trade - 161-163
—— on shipping - - - 170
—— on trade - - - 151, 152
—— proposals concerning provision trade - - 254-256
—— sends weavers from Brabant, &c. - - - - 189
—— statistics of damage caused by the Rebellion - - - - - 109
Tenant-right - - - 14, 25, 30
—— —— allowed by Cromwellians - - - - 134
—— —— attacked - - - 218
Tenants mainly natives 25-30, 133
Tenure, see Land-tenure.
Thomond (Earl of), pres. of Munster: his house at Bunrattie - - - - - 54
Tillage, decay of - - - 96
—— extent of - 41-43, 144-146
—— hampered by land system 235
—— hampered by military - 212
—— inferior - - - - 37
Timber exports - - - - 43
—— see generally under Woods.
Tipperary silver and lead mines - - - - 53
Tithes and tithe war - - 121
Tobacco, additional duty on - 234
—— consumption of - - 139, 140
—— import duty on - - - 198
—— monopoly - - - 60, 65
Tokens, tradesmens' - 115, 204
Tolls, excessive - - - - 60
Tomond iron works - - 51, 52
Tonnage and poundage, Acts of - - - - - 150, 199
—— described - - - - 198
Toome smith famous - - 49
Tories, otherwise Kerns, Rapparees, Capt. Rocks, and Wood-kerns 25, 26, 31, 34, 133, 134
Towns decayed - - - - 112
—— and houses destroyed - 213
—— exclusion of natives from - - - 112, 113
Trade - - - 57-68, 150-173
—— adverse balance of - - 206
—— little foreign - - - 139
—— little home - - - - 175
—— oppressed by England 220, 232
—— ruined by the Rebellion - 110
—— under Elizabeth and the Stuarts - - - - 86
Trade-Act, Patriot Parliament's - - 171-173, 261-264
Transplantation of Catholic landowners - - - 129
—— cottiers and labourers exempted from - - 136
Transplantations, Cromwellian - - - - - 35
Transportations to the colonies - - - - 102, 214
Tredagh, no woods near - - 46
Turf used as fuel 44, 48, 139, 148
Turks harass fisheries - - 82
Twentieth parts, the - - 87
Tyrconnell (Lord) fosters trade - - - - - 169

U.

Ulster, plantation of 22, 27, 33, 42, 43, 78
—— tenant-right in - - - 134
—— woods - - - - - 47
—— woods wasted - - - 43
—— linen industry thrives 192, 231
—— only one tanner in - - 85
Uniformity, Act of - - - 13
Upper Ossory (Baron of): his grants - - - - 35
Usquebaugh drunk - - - 15
—— manufacture and export 84
—— see also Aqua vitae.

V.

Venice, export of fish to - - 195
Venetian glass industry in Munster - - - - 83

INDEX.

	PAGE
Venomous beasts none	141
Victualling of ships	163, 164
Virginia, transportations to	24

W.

	PAGE
Wages in 1608 and 1640 (App. II.)	242, 243
—— agricultural labourers', low	36
Wandsworth's (Christopher) iron mines	48, 49
War, Franco-Spanish, injures Irish trade	168
—— Spanish, injures Irish trade	61
Wards and liveries, courts of, established	91
Waterford, arms issued from	106
—— corn imported into	105
—— woollen industry	186
Wax, manufacture of	197
Weavers exempted from Juries	188
—— immigration of English, to Ireland	228
—— immigration of French	79
—— —— of Jersey	189
—— settlement of foreign, encouraged	189
"Weeds," the Irish so called in Limerick	27
Weights and measures not fixed	61
—— for weighing foreign coins	206
Wentworth (Sir Thomas) and Irish carrying trade	67
—— and the linen industry	76
—— anxious to increase Irish revenue	63, 64
—— develops national resources	222
—— fosters Irish trade	63, 64
—— grants monopolies	60
—— improves commerce	8, 9
—— on Irish trade	9, 10
—— policy criticized	11
—— proposes salt monopoly (App. III.)	244, 245
—— reforms the revenue	92
West India trade, embargo on	169
Wexford, iron works near	149
—— plantations	23-25
—— population of	34, 35
—— shipping	67
—— want in	107
—— Co., woods in	47
Wheat, introduction of, by English	41
—— prices in 1643	105
see also Corn.	
Whiskey, manufacture of	144
see also Aqua Vitae, &c.	
Whitty (Adam) of Arklow, obtains patent to make glass	83
Wicklow, Co., woods in	47
"Wild Geese"	214
William III. favourable to Irish trade	219

	PAGE
Windows scarce	141
Wine, duties on	65, 87, 198
—— monopoly oppressive	60
Wolves exterminated	184
—— numerous	108
Wood-kerns, see under Tories.	
Woods, the	43-47, 147, 148
—— destroyed	67, 106, 213
Wool exchanged for munitions	70
—— export duties on	69
—— export of	75, 160, 162, 187
—— export of unlicensed, felony	180, 181
—— export prohibited	58, 69
—— export to England	64, 65, 178-180
—— fees, &c., on (Table)	179
—— license duty	199
—— licenses to export	74, 177, 178
—— of good quality	184
—— a principle product	178
—— Sheridan's proposal for buying	265-268
—— smuggled abroad	118, 164, 180
—— smuggled to Spain	72
—— smuggling connived at	178
—— smuggling much practised	73
Woollen and linen industries compared	230-231
—— goods, their export forbidden	214
—— industry	68-75, 176-187
—— industry attacked	221, 225, 226, 228
—— —— criticized by Cary	226
—— —— encouraged	156, 174
—— —— petitioned against	228
—— —— prosperous	227
—— —— revived under Grattan	230
—— manufacture at Callan	268, 269
—— —— suppressed	234-236
—— manufactures, export forbidden	229
—— —— restrained	10
—— trade in England suffers	165
—— —— increases	166
—— —— regulated by statute	199
—— —— stimulated	118
Woollens, English protective import duty on	152, 229
—— exports (Table)	227
—— prohibited	234
—— imports (Table)	228
—— increased export duty on	234
—— of high reputation abroad	227
Wool-staple, a, proposed	181
Woolworks numerous in 1672	184

Y.

	PAGE
Yarn exported to England	58
—— homespun	138
Yeomen farmers of England	219
Youghal, fishing off	195
—— houses destroyed	112
—— linen manufactures	78
—— staple established at	72

88198